Somatic and Autonomic Regulation in Sleep

W0043025

Springer
Milano
Berlin
Heidelberg
New York
Barcelona
Budapest
Hong Kong
London
Paris
Santa Clara
Singapore
Tokyo

E. Lugaresi • P. L. Parmeggiani (Eds)

Somatic and Autonomic Regulation in Sleep

Physiological and Clinical Aspects

 Springer

PROF. ELIO LUGARESI
Università di Bologna
Istituto di Neurologia
Via Ugo Foscolo 7
40123 Bologna, Italy

PROF. PIER LUIGI PARMEGGIANI
Università di Bologna
Dipartimento di Fisiologia
Umana e Generale
Piazza Porta San Donato 2
40127 Bologna, Italy

Die Deutsche Bibliothek - CIP-Einheitsaufnahme
Somatic and autonomic regulation in sleep : physiological and
clinical aspects / E. Lugaresi ; P. L. Parmeggiani (ed.). - Milano ;
Berlin ; Heidelberg ; New York ; Barcelona ; Budapest ; Hong Kong ;
London ; Paris; Santa Clara ; Singapore ; Tokyo : Springer, 1997
ISBN 978-88-470-2277-5   ISBN 978-88-470-2275-1 (eBook)
DOI 10.1007/978-88-470-2275-1

Fotocomposizione: Graphostudio, Milano

SPIN: 10629610

Preface

This volume contains the contributions to a course entitled "Sleep and its Pathology", which was organized by the Advanced School of the Italian Neuroscience Society at the "Alessandro Volta" Center of Scientific Culture (Villa Olmo, Como, Italy, May 9-10, 1996).

The course was aimed at informing the medical audience about recent developments in the field with particular regard to the work of the sleep laboratories of the University of Bologna.

The first part of this book presents experimental results dealing with the biochemical specificity of hypothalamic sleep mechanisms, cerebral metabolism during sleep and the sleep-dependent systemic cardiovascular adjustments in relation to blood perfusion and thermal homeostasis of the brain. The second part covers pathophysiological aspects of human sleep, namely the circadian rhythm of body core temperature in neurodegenerative diseases, the descriptive epidemiology of excessive daytime sleepiness, the disorders of breathing and motor control in sleep and the syndrome of nocturnal frontal lobe epilepsy. The Editors hope that the book may be useful not only to specifically interested readers, but also to general practitioners.

The Editors wish to express special thanks to Professor Eugenio E. Müller for his suggestion to collect the lectures in book-form.

The Editors thank the authors for their contributions to the course and to this publication and express their appreciation to Springer-Verlag for helping make this monography possible.

Bologna, Juli 1997
Elio Lugaresi
Pier Luigi Parmeggiani

The Authors wish to thank Synthélabo for the support
in the realization of this volume

Indice

Part I
Biochemistry, Metabolism and Physiology of Sleep

Part II
Pathophysiological Aspects of Human Sleep

List of Contributors

AMICI R., 3

AZZARONI A., 43

CERULLO A., 125

CIRIGNOTTA F., 87

COCCAGNA G., 87

CORTELLI P., 55

D'ALESSANDRO R., 73

FRANZINI C., 25

LENZI P., 25

LUGARESI E., 55, 107, 125

MONTAGNA P., 107

PARMEGGIANI P.L., 43

PEREZ E., 3

PIERANGELI G., 55

PLAZZI G., 55, 107, 125

PROVINI F., 55, 107, 125

RINALDI R., 73

TINUPER P., 125

TONON C., 73

VIGNATELLI L., 73

ZAMBONI G., 3

ZOCCOLI G., 25

List of Abbreviations

ACTH   adrenocorticotropic hormone
ADH   antidiuretic hormone
ADNFLE autosomal dominant nocturnal frontal lobe epilepsy
AF   autonomic failure
AI   apnea index
ALS   amyotrophic lateral sclerosis
cAMP   adenosine 3':5'-cyclic monophosphate
ANS   autonomic nervous system
AS   active sleep

BcT   body core temperature
BiPAP   bilevel positive airway pressure
BMI   body mass index
BNSQ   basic nordic sleep questionnaire

CAN   central autonomic network
CBF   cerebral blood flow
COPD   chronic obstructive pulmonary disease
CPAP   continuous positive airway pressure
CR   circadian rhythm

ECG   electrocardiogram
EEG   electroencephalogram
EDS   excessive daytime sleepiness
EMG   electromyogram
EOG   electro-oculogram
ESS   Epworth sleepiness scale

FFI   fatal familial insomnia
c-fos   cellular-FBJ (Finkel, Biskis, Jinkins) murine osteogenic sarcoma virus oncogene (proto-oncogene)

GH   growth hormone
GHT   geniculohypothalamic tract

HF   high frequency
HR   heart rate

IGL   intergeniculate
IPPV   intermittent positive pressure ventilation

Km   Michaelis constant

LF   low frequency band
LH   luteinizing hormone

MSA   multiple system atrophy
MSLT   multiple sleep latency test

NM   nocturnal myoclonus
NREM   Non REM or synchronized sleep

OSAS   obstructive sleep apnea syndrome

PAF   pure autonomic failure
PaCO$_2$ partial pressure of carbon dioxide in arterial blood
PaO$_2$ partial pressure of oxygen in arterial blood
PO-AH   preoptic-anterior hypothalamic area
PRL   prolactin
PRNP   prion protein
PSM   propriospinal myoclonus
PSQI   Pittsburg sleep quality index

QS   quiet sleep

RBD	REM sleep behaviour disorders	Thy	hypothalamic temperature
REM	rapid eye movement or desynchronized sleep	Tp	pons temperature
		TSH	thyroid-stimulating hormone
RHT	retinohypothalamic tract		
RLS	restless legs syndrome	UPPP	uvulopalatopharyngoplasty
SaO$_2$	percent saturation of hemoglobin with oxygen in arterial blood	Vmax	maximal rate of formation of products in an enzyme catalyzed reaction
SCN	suprachiasmatic nucleus		
SWAI	sleep-wake activity inventory		
Ta	ambient temperature	W	wake

Part I

Biochemistry, Metabolism and Physiology of Sleep

Biochemical Approach to the Wake-Sleep Cycle

G. Zamboni, E. Perez and R. Amici

Introduction

The biochemistry of waking and sleeping processes seems to engulf the whole of the brain's functions; the molecular description of the mechanisms involved constitutes one of the most fundamental explanatory levels available. In the case of the wake-sleep cycle, the gap in knowledge existing between molecules and these behavioural states has been filled on the assumption that wake-sleep processes are the expression of some general function concerning both the brain and the body, and thus, both widespread neuronal and humoral influences have been investigated. Both pathways imply that neuronal or non-neuronal cells release substances which control the behavioural states of waking and sleeping by acting on regulatory centres. However, there are two tasks that should be fulfilled in order to clarify how the control of waking and sleeping is organized at the molecular level: the first is to identify the substance(s) involved and the other is to assess the location(s) of their effects, that is, the components of the nervous circuit involved in the control.

Basically, the approaches used in the past have been to increase the signal-to-noise ratio in the analysis of specific brain chemicals by enhancing the release of sleep-inducing substances or *sleep factors*, mainly by means of sleep deprivation [1], or by manipulating the activity of neuronal systems with diffuse projections to the brain [2].

In the first instance, following the identification of the chemical nature of the sleep factor, animals are used as a bioassay system to evaluate the effects of the administration of the substance. As in any classical pharmacological approach, the problem is to assess the specificity of the effects observed. In fact, the picture which seems to emerge from early [1] and more recent studies [3] is that the specificity of sleep factors is questionable, since these substances appear to regulate many other functions. Another way to address this question is to consider that, since wake-sleep processes represent particular modalities of physiological regulation [4-6], it is highly probable that many substances which control body functions might also influence sleep.

Dipartimento di Fisiologia Umana e Generale, Università di Bologna, Piazza di Porta S. Donato 2, 40127 Bologna, Italy

In the second instance, studies originated from the findings that stimulation or lesion of the brainstem reticular formation changed the pattern of the bioelectrical cortical activity (see [7]). It is impossible to review here the vast literature concerning the role of the brainstem in wake-sleep processes. However, a summarized perspective may be attempted on the basis of the relationship between wake-sleep and physiological regulations. The brainstem contains cell groups involved in the regulation of autonomic functions (e.g. cardiorespiratory control), which have a close relationship with the visceral sensory relay nuclei of the brainstem, such as the nucleus of tractus solitarius, the nucleus parabrachialis and the locus coeruleus, and with the integrative higher centres of autonomic activity, such as the hypothalamus and the amygdala [8-10]. From an anatomical point of view, the nucleus of the tractus solitarius, the nucleus parabrachialis, and other cell groups like the periacqueductal gray and the dorsal motor nucleus of the vagus are connected to the hypothalamus and the preoptic region via the fasciculus longitudinalis dorsalis and the median forebrain bundle. This system is particularly rich in neurotransmitters and neuromodulators, and has been considered to constitute a unit named the *paracrine core of the neuraxis* [11,12] on the basis of its chemical architecture and its involvement in the control of homeostasis and reproductive activity [11-14]. The paracrine core has two adjuncts which are continuous with the core region: the raphe nuclei on one side; and the locus coeruleus, other noradrenergic brainstem groups and the pontomesencephalic tegmentum on the other [11, 12, 15-18].

It is immediately evident that such a complex functional system contains most of the brain regions that are involved in the control of the wake-sleep cycle [2, 4, 6, 19]. Thus, it appears that the relationship between wake-sleep processes and physiological regulation rests on a structural framework encompassing several interconnected networks. On this basis, it is apparent that an investigation concerning wake-sleep processes carried out at the molecular level should take into account the involvement of different neural substances and regions.

One possibility to simplify the molecular analysis of behavioural states is to take into account the fact that the chemical complexity of the brain, at the communication level, may converge on a very low number of central metabolic pathways. Thus, the study of brain intermediate metabolism might shed light on the molecular description of wake-sleep processes on the basis of the brain energy flow. Intermediate brain metabolism during sleep has been prevalently studied on the whole brain in the past [20, 21], and more recently, on a regional basis by using radioactive glucose analogues that are trapped within cells upon phosphorylation [22-24].

Such an approach is facilitated by the fact that the brain uses only the oxidation of glucose for its metabolism in most normal conditions. It should be considered, however, that there is about 20% excess in the glucose utilization, with respect to oxygen consumption, which is not explained by a relative lack of oxygen caused by other oxidative processes such as those carried out by monoamine oxygenases or by monoamine oxidases [25]. In fact, it has been observed that brain non-oxidative glucose consumption may also increase with focal neural activity

[26]. Attempts to clarify the fate of the metabolic energy derived from glucose utilization in nervous tissue have been sparse, but some observations suggest that it is primarily utilized to maintain or restore ionic gradients [27] or for synaptic activity [23, 28].

The advantage of using radioactive glucose analogues is that it is possible to draw a metabolic map of brain activity by autoradiography or positron emission tomography. The main limitation of this method, as far as the wake-sleep cycle is concerned, is the duration of the chase period which follows the pulse of radioactive precursor, since in some cases it may be longer than the average duration of the shortest sleep episode, that is, REM (rapid eye movement) sleep. This problem has been approached in different ways [29-33].

In short, such metabolic maps [23, 24] indicate that changes in glucose utilization involve several brain regions, strongly supporting the idea that wake-sleep regulation is based on a network of brain structures functionally interconnected. However, it also appears that these data do not yet define a clear relationship between the changes in glucose metabolism and the wealth of functional results obtained with other techniques. Albeit that a state-dependent change in glucose utilization might not always be related to functionally relevant changes in the state-regulation, the attempt to match functional and metabolic data may still be considered a useful approach which sets some functional limits for the areas involved in wake-sleep processes.

Although glucose is the obligatory fuel of the brain, a regulating neuronal network might carry out some relevant activities without noticeable metabolic changes. From this viewpoint, the approach we would like to review here is the investigation of one of the first intracellular step of interneuronal chemical communication: changes in the concentration of the second messenger adenosine 3':5'-cyclic monophosphate (cAMP). In the brain, as in other tissues, both cAMP and cAMP-dependent protein phosphorylation represent important convergence pathways of many different extracellular regulatory signals [34, 35]. The experimental relevance of studying a second messenger such as cAMP is not only related to its intracellular targets which generate the biological responses, i.e. enzymes, transcription factors or ion channels, but also to the molecular amplification of extracellular, i.e. first messenger, signals which stimulate its biosynthesis.

Thus, the approach of studying changes in cAMP levels with respect to waking and sleeping is an additional attempt to simplify the molecular analysis of brain activity at the communication level. With respect to the regional determination of metabolic activity by means of glucose utilization, it has the advantage of better molecular resolution, with the disadvantage in that it is difficult to draw precise maps of changes.

For this reason, the biochemical studies reviewed here concentrate on one of the more relevant knots of the neural network subserving wake-sleep regulation, the preoptic-anterior hypothalamic area (PO-AH). A great wealth of data [4, 36, 37] has firmly established the role of this area in the regulation of the wake-sleep cycle in mammals on the ground of autonomic regulations or, in other words, of homeostasis. Moreover, these studies have also shown that environmental influences on sleep may be understood on a physiological basis, and that ambient conditions may be changed to modify wake-sleep processes within a common interpretative frame.

Since we have used this frame of reference in our approach, in that animals (albino rats) have been exposed to both normal laboratory and low ambient temperature (Ta), a discussion of the principles and the facts relevant to our model will be given below.

The Wake-Sleep Cycle of the Rat at Different Ambient Temperatures in the Frame of Physiological Regulations

General Remarks

The interaction between wake-sleep and physiological regulations was first observed in the cat in the form of an impairment of thermoregulatory control in REM sleep, interpreted as the consequence of a change in the hypothalamic integrative activity [38]. There is not yet a satisfactory interpretation of the impairment of thermoregulatory control during REM sleep. However, it appears that the most productive approach is to utilize the heuristic power of phenomenal generalization. On this ground, wake-sleep processes have been considered on the basis of both phylogenesis [39] which takes into account the supraorganismic level of their organization, and of physiological homeostasis [4, 5] which takes into account the organismic organizational level. Phylogenetic data indicate that the wake-sleep cycle has evolved in mammals in association with endothermy. Physiological homeostasis shows that the wake-sleep cycle is characterized by a functional dichotomy in which wakefulness and non-REM (NREM) sleep are stages with fully operant homeostatic controls (albeit at different levels), and REM sleep is a stage in which homeostatic controls are impaired.

It may be possible to unify both viewpoints either on a structural or on a functional ground. In the first instance, the role of the hypothalamus and, in particular the PO-AH, in thermoregulatory control in mammals [40, 41] may be considered an evolutionary-driven specialization of other functions, for example the control of volume and composition of body fluid compartments and endocrine activity, already controlled by the hypothalamic structures in lower vertebrates [42, 43]. In the second instance, the autonomic changes that characterize the wake-sleep cycle may be considered as qualitative and quantitative equivalents of other kinds of autonomic regulations, such as those related to physical activity. In other words, since highly integrated homeostatic controls involve effectors subserving different functions, the stages of the wake-sleep cycle should be matched with other physiological regulations according to a set of priority rules, shaped by the evolutionary adaptation to ambient constraints. This process, named homeostatic competition [44], should reach its clearest expression in the regulation of REM sleep in homeotherms, since the suspension of the hypothalamic integrative activity observed during this stage affects thermoregulation more than it does the other autonomic functions (for example those related to cardiorespiratory control) which rely on a higher degree of autonomy from diencephalic control [36].

In spite of the early observations of the influence of Ta on sleep [45, 46], few animal studies have investigated the wake-sleep pattern during exposure to different ambient temperatures, and these experiments differ in both the temperature and the duration of the period of exposure (see [47]). These studies indicate that sleep modifications depend initially on the direction of change of Ta, with respect to the temperature of adaptation (normal laboratory Ta). For example, when Ta was raised only a few degrees above the normal laboratory value, both REM sleep and NREM sleep tended to increase, whilst when Ta was lowered to below control values, REM and NREM decreased. However, the occurrence of REM sleep dissociated from that of NREM sleep, decreasing at a faster rate when the divergence of Ta from the normal laboratory values deepened.

The fact that the body temperature of mammals lies in the upper part of the respective range of physiologically tolerated ambient temperatures makes the span of the intervention of cold defense mechanisms much wider than that of heat defense mechanisms. Thus, it would appear that the functional relationship between the periods of homeostatic regulation during the wake-sleep cycle, that is in wakefulness and NREM sleep, and those of non-homeostatic regulations, that is REM sleep, might be modulated more precisely at low rather than at high ambient temperatures in proportion to the thermal load and to the homeostatic regulation of sleep.

The concept of homeostatic regulation of sleep rests on the assumption that a feedback regulatory mechanism compensates for gains and losses by changing sleep propensity in accordance with the deviations from a reference level [39, 48, 49]. On a broad basis, this concept is underlied by sleep deprivation and recovery, or sleep rebound, which ensues if the deprivation has reached a certain level. For NREM sleep, the expression of the intensity may be represented by electroencephalogram (EEG) slow wave activity (mean power density in the range of 0.75-4.0 Hz), which is taken as a measure of underlying specific processes [49]. In the rat, which is a nocturnal animal, slow wave activity is higher at the beginning of the light period and declines during the following hours of sleep [50, 51]. Moreover, slow wave activity is increased following sleep deprivation [52, 53] and it has been proposed that, in some instances, a deficit of NREM sleep is compensated for by an enhancement of sleep intensity without any increase in the duration [54]. In contrast, REM sleep does not seem to offer such a clear-cut electroencephalographic equivalent of intensity, although a prolonged increase in the theta activity (mean power density in the range of 6.25-9.0 Hz) has been observed following deprivation [54, 55].

The lack of an electroencephalographic parameter for intensity suggests that REM sleep processes may not be represented on the frequency domain of nervous bioelectrical activity. From the suspension of homeostatic regulations during REM sleep, it may be inferred that the temporal boundaries of episodes are a characteristic which attains to the physiological control of the organism. Thus, the temporal analysis of REM sleep, that is the determination of the length of the episodes and of their intervening intervals, should provide either a direct measure of the intensity of the specific processes related to this stage, or an indirect evaluation of the engagement of homeostatic regulations. This should be particularly relevant for any analysis carried out on the nervous structures involved in the control of homeostasis.

Experimental Evidence

Figure 1 shows the distribution of intervals separating consecutive REM sleep episodes, that is the time from the end of one REM sleep episode to the beginning of the next, during the wake-sleep cycle of rats kept at normal laboratory temperature. We have set the threshold duration for each REM sleep episode and interval

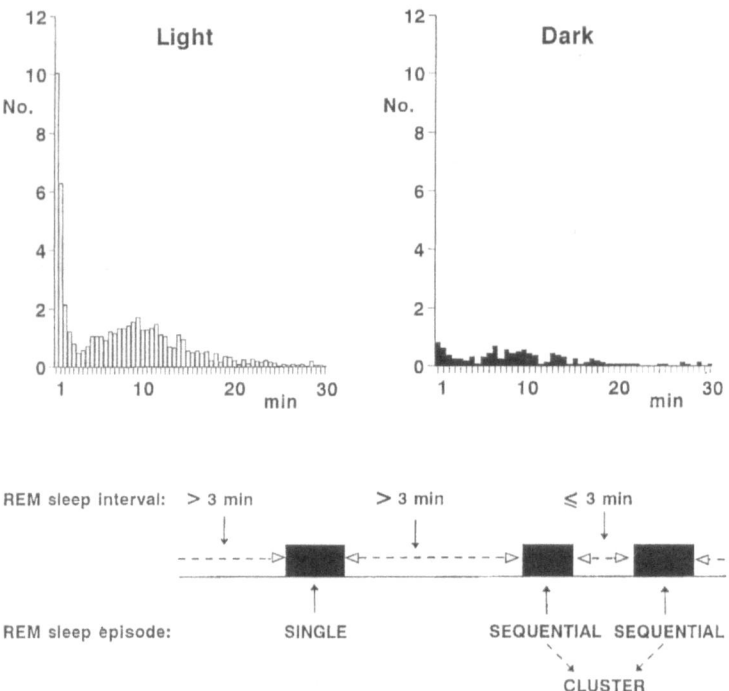

Fig.1. *Frequency distribution of REM sleep intervals and partition of REM sleep episodes. Top* Frequency distribution of the time interval from the end of one REM sleep episode to the beginning of the next (REM sleep interval), during either the light (*left*) or the dark (*right*) period of the light-dark cycle, in rats kept under normal laboratory conditions (12 h:12 h light-dark cycle; Ta 23°C). The width of the class intervals is 30 s and only data relative to the first 60 classes are shown. Each bar represents the mean class value. *Bottom* Partition of REM sleep episodes into single and sequential episodes on the basis of the bimodal distribution of their intervals. The single REM sleep episodes are defined as those which are both preceded and followed by long REM sleep intervals (>3 min), whilst the sequential REM sleep episodes are those which are separated by short (≤3 min) intervals and are found in clusters. Data are from rats kept under normal laboratory conditions

at 10 s, on the basis of both electroencephalographic and behavioural criteria: an interruption of REM sleep has been considered as such only when the disappearance of the theta rhythm from the EEG trace was concomitant with a movement of the animal, objectively recorded by means of a passive infrared detector. The distribution of REM sleep intervals follows a circadian pattern, and is clearly bimodal in the rest (light) period of the rest-activity cycle. The bimodal distribution identifies two populations of intervals which can be empirically divided into short and long intervals, by setting the boundary between them at the minimal frequency class. Thus, short intervals are ≤ 3 min and long intervals > 3 min. Figure 1 shows that short and long REM sleep intervals may also be used to subdivide REM sleep episodes which, on this basis, can be divided into single REM sleep episodes when they are separated by long intervals, and sequential REM sleep episodes when they are separated by short intervals [56]. Sequential REM sleep episodes occur as clusters, in which the first and last units are respectively separated from the preceding and the following REM sleep episodes by long intervals.

It is obvious that the distribution of REM sleep intervals depends on the criteria used in the definition of the minimal duration of both the interval and the REM sleep episode. This problem has been addressed in an analysis of the length of the REM-NREM sleep cycle, that is of the interval between the onset of consecutive REM sleep episodes, in the rat [57]. The cycle duration falls in the normal range for this species (7-13 min) only when the threshold duration of both REM sleep episodes and intervals is between 20 and 60 s. Significant differences in the duration and number of cycles between the normal condition and that of sleep deprivation (by forced locomotion) is evident only for threshold durations close to the higher value of the above temporal range. The REM-NREM sleep cycle constitutes one of the possible ultradian representations of wake-sleep processes with respect to the completion of the cycle. However, by taking in consideration the sum of the extensions of the two physiologically different periods (REM vs. NREM sleep and wakefulness), it overlooks the temporal aspect of their functional relationship. On this ground, it is worth mentioning that a bimodal distribution of the intervals between REM sleep episodes has been observed in other species, such as monkey [58], cat [59] and man [60, 61].

Figure 2 shows how the amount of REM sleep, in the form of both single and sequential episodes, changes as a function of the sum of the preceding and following intervals. These diagrams depict REM sleep as a period of discontinuity occurring within a larger time interval, during which physiological variables are kept in their normal range. It appears that the amount of REM sleep, either in the form of single or sequential episodes, increases with the duration of the surrounding periods according to a curve which reduces progressively in its steepness. However, at any given duration of the surrounding (homeostatic) periods, the amount of REM sleep in the form of sequential episodes is always substantially higher than that in the form of single episodes. Therefore, it should be expected that a shift from the curve concerning single REM sleep episodes to that concerning sequential REM sleep episodes would be under tighter control than that

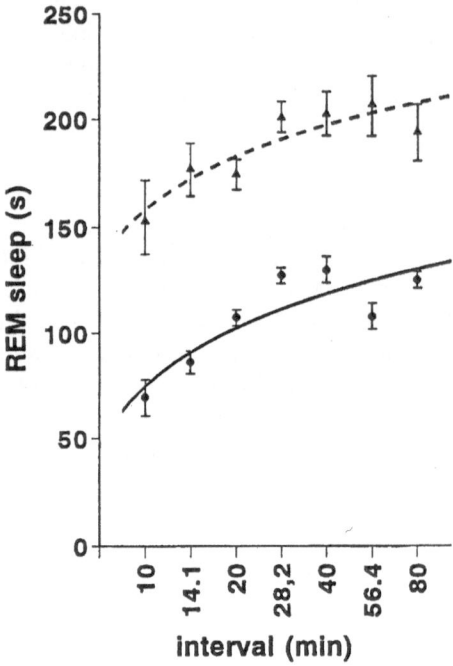

Fig. 2. *Relationship between the cumulative duration (min) of the REM sleep intervals which preceded and followed either a single REM sleep episode (solid line) or a REM sleep cluster (broken line), and the mean duration (s) of either the single REM sleep episode or the REM sleep cluster which occurred within those two REM sleep intervals.* The duration of the short REM sleep interval(s) within each REM sleep cluster has been cumulated to that of the two REM sleep intervals. The midpoint of each class is shown on a logarithmic scale. Data are from rats kept under normal laboratory conditions

which may be actuated along each of the curves. Thus, from the viewpoint of physiological regulations, sequential REM sleep episodes occur in an inverse proportion to the intensity of any contingent homeostatic conflict.

The diagrams of Fig. 3 represent time courses for the amount of REM sleep, in the form of both single and sequential episodes, in different groups of rats during normal laboratory conditions, after 24 h and 48 h exposures to two low ambient temperatures, and after a subsequent 12 h period of recovery, carried out by bringing animals back to the normal laboratory temperature. Detailed analysis of REM sleep occurrence in these conditions showed that the amount of this sleep stage changes by means of a modification of the frequency and not of the average duration of episodes [56, 62]. The only notable exception is represented by the first 12 h period at either Ta 0°C or Ta -10°C, during which the average duration of REM sleep episodes was significantly reduced. However, the duration of episodes observed in normal conditions was reattained immediately afterwords. The tendency to change the amount of REM sleep, by changing the frequency and not the duration of episodes, has been observed at different ambient temperatures in

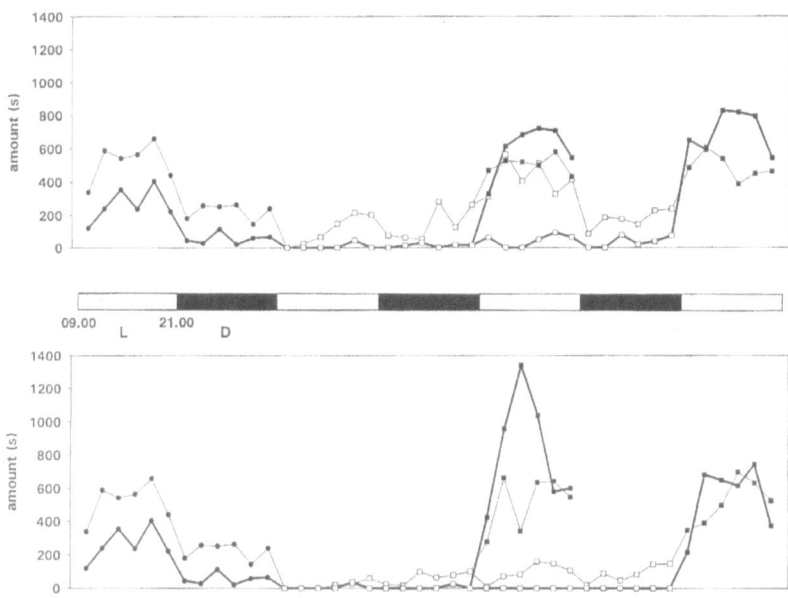

Fig. 3. *Time course of REM sleep in the rat in the form of either single REM sleep episodes (thin line) or sequential REM sleep episodes (thick line).* Data points represent cumulative seconds of REM sleep in 2 h periods throughout the experiment. *Top left to right* 24 h period under normal laboratory conditions (12 h:12 h light-dark cycle, Ta 23°C, *filled circles*); 24 h/48 h exposure to Ta 0°C (*unfilled squares*); 12 h recovery at Ta 23°C which followed either 24 or 48 h exposure to Ta 0°C (*filled squares*). *Bottom left to right* 24 h under normal laboratory conditions (*filled circles*); 24 h/48 h exposure to Ta -10°C (*unfilled squares*); 12 h recovery at Ta 23°C which followed either 24 or 48 h exposure to Ta -10°C (*filled squares*). Data calculated from the results of [56, 62]

other mammals of very small size [63-65]. This suggests that in species with an unfavourable surface-to-volume ratio a tight control on thermoregulation sets an upper limit to the duration of REM sleep, which can only be overrun by adding distinct episodes in close vicinity, that is by introducing some short periods of homeostatic control to check a long lasting gap in homeostatic regulations. However, since the tendency to keep the duration of REM sleep episodes constant is also present at low ambient temperature, these results underlie the importance that REM sleep might have for mammals. In fact, the drive for REM sleep may be suppressed only when the ambient condition is very challenging; this has been first shown in the cat [45, 46].

As far as the amount of REM sleep is concerned, Fig. 3 shows that at normal laboratory temperature both single and sequential REM sleep episodes follow a circadian pattern. This pattern is altered by the exposure to low Ta since the occurrence of REM sleep is immediately depressed. However, following the first 24 h period, such an alteration is directly related to the thermal load experienced by the animals, that is with either the ambient temperature or the duration of the exposure.

At Ta 0°C, the circadian pattern in the occurrence of single REM sleep episodes is recovered in the second 24 h period of exposure. Both the light and the dark amounts of REM sleep, in the form of single episodes, lie close to 80% of the control value. Also, the amount of REM sleep in the form of sequential episodes is kept at a negligible level for the whole exposure period. Thus, in terms of the relationship represented in Fig. 2, it appears that the ambient condition allows both the circadian and the REM sleep homeostasis factors, which are able to affect both forms of REM sleep episode at normal laboratory temperature, to operate only on the lower curve related to single episodes. The weight of the ambient condition is clearly shown by the fact that, in the second 24 h period at Ta -10°C, single REM sleep episodes were depressed as strongly as sequential episodes.

The recovery periods following the exposure to both ambient temperatures show the existence of a rebound of REM sleep. This result confirms that the homeostatic control of this stage of the wake-sleep cycle may be represented on the temporal domain of the duration of its episodes. With respect to this, the study of the relationship between the loss and the recovery of REM sleep in the cat has led to the precise determination of the hourly rate of accumulation of the REM sleep debt in this species [66]. As far as the rat is concerned, the results in Fig. 3 indicate that the normal reappearance of single REM sleep episodes during the second 24 h period at Ta 0°C is driven by sleep loss in the first 24 h period. The balance between REM sleep and thermoregulatory homeostatic controls suggests that the quick adaptation of the rat to Ta 0°C may be a further sign of the relevance of REM sleep for mammals. However, these results also point to two particular characteristics of the recovery. First, the rebound is due to a change in the amount of REM sleep in the form of sequential episodes, whilst that in the form of single episodes is kept at the same level observed at normal laboratory temperature. Second, whilst both the rebounds following the 24 h exposure to low Tas and following the 48 h exposure to Ta 0°C are broadly proportional to the amount of the sleep loss, that following the 48 h exposure to Ta -10°C is not. The importance of this finding may be again discussed in light of Fig. 2. The recovery condition is characterized by a return to normal laboratory Ta, and thus, by the abolition of a strong constraint to shift to the sequential REM sleep episode curve. Such a shift allows REM sleep to be produced in a more efficient way and in relation to the amount lost during the exposure. However, a different type of constraint appears to be present with respect to the second 24 h period spent at Ta -10°C. This suggests that the nervous structures regulating the occurrence of sequential REM sleep episodes have been affected by the extension of the period of exposure to the lowest ambient temperature.

It is evident that, on the basis of the previous discussion and on these results, the brain region most likely to be involved in such a response is the preoptic-anterior hypothalamic area. The biochemical evidence of such an involvement is discussed further.

cAMP Changes in the Preoptic-Anterior Hypothalamic Area with Respect to the Wake-Sleep Cycle

General Remarks

The list of neuroactive substances, that is of chemicals acting as neurotransmitters or first messengers in the brain, comprises not less than fifty different items. Basically, they range from hormones and autacoids to classical or peptidergic neurotransmitters released or co-released during synaptic activity. Many of these substances act on receptors linked to adenylate cyclases by means of membrane-associated proteins acting as signal transducers (G proteins), and affect the accumulation of cAMP (for a general review see [67]). However, other signalling mechanisms that influence intracellular calcium ion concentration may also regulate the activity of adenylate cyclases [68]. In the brain, as in many other tissues, adenylate cyclase exists in many isoforms which appear to be differentially regulated and expressed in diverse regions [69].

Following stimulation by first messengers, the normal cellular cAMP concentration may undergo a manifold rise in a matter of seconds [70]. However, the maintainance of cellular cAMP concentration depends on the relative rates of synthesis by adenylate cyclases and degradation by cyclic nucleotide phosphodiesterases. These latter enzymes are regulated by many of the factors that affect adenylate cyclases and also exist in many isoforms, selectively expressed in different brain regions and tissues [71].

cAMP, an allosteric regulator of cellular activity, has as its main target a set of cAMP-dependent protein kinases [72]. Also, the finding that cAMP may directly gate some ion channels [73] is relevant to nervous activity. Protein kinase is the starting point of a widely divergent pathway of reversible, covalent modifications of intracellular proteins. In analogy with the maintainance of intracellular cAMP levels, the protein phosphorylation state also depends on dephosphorylating enzymes known as protein phosphatases [74]. The balance between protein kinase and phosphatase activity is the intracellular regulatory step of receptor-mediated signalling pathways.

From a biochemical point of view, the presence of several cycles of enzymatic activity in the cellular transduction of extracellular signals may allow more refined controls and greater amplifications. However, as far as nervous activity is concerned, it should be noted that these intracellular events translate the effects of synaptic activity on a temporal domain which is much wider than that normally used by chemically-gated ion channels. In fact slow bioelectrical synaptic responses are dependent on receptors coupled to the G protein transducing system [34], and the structural changes characterizing synaptic plasticity (related to some forms of learning) are also dependent on second messengers systems [75].

It then appears that the temporal domain of cellular events mediated by second messengers is of the same order of magnitude not only of the events which characterize the wake-sleep cycle, but also of the processes which underlie sleep homeostasis, since for example, it is likely that both sleep deprivation (accumulation of a specific sleep debt) and recovery (storage of some specific memory trace) may be two behavioural aspects of the same structural change. On this ground, it is interesting to note that some of the activity of thalamocortical neurons involved in the genesis of wake-sleep related bioelectrical rhythms is under control of G protein coupled receptors [76].

Experimental Evidence

The common feature in these experiments is that cAMP concentration was determined following the sacrifice of animals in liquid nitrogen. The cage floor was opened by remote control when the sacrifice was made under electroencephalographic control. Although liquid nitrogen blocks brain enzyme activity with a delay a few tens of seconds longer than the fastest method available (rapid heating of the tissue by microwaves), it has the advantage of preserving better the fine anatomy of the brain and avoiding the need to restrain the animals to keep the head within an electromagnetic field. We have overcome the obvious disadvantage of the dissection of the frozen tissue by developing a technique in which the brain was fragmented according to the natural cleavage planes passing through the cerebral commissures. Samples from the different regions were taken under stereomicroscopic control by using small stainless steel needle punches with the tip cut at 90° and sharpened. All the operations were performed in a box kept at subzero temperature by means of dry ice.

The results shown concern only the preoptic anterior hypothalamic area (PO-AH). In all experiments, control samples were taken from the cerebral cortex, since wake-sleep stages are classified according to cerebral cortical activity. However, no significant changes in cortical cAMP concentration were found in the experimental conditions outlined below.

Figure 4 shows the changes in the PO-AH cAMP concentration found during the different stages of the wake-sleep cycle in the light and dark hours of the normal 12 h:12 h light-dark cycle. In this experiment, animals were sacrificed during the interval between one-half and two-thirds of each light-dark period. Progression of the wake-sleep cycle from wakefulness to REM sleep was marked by a progressive decrease in the cAMP concentration. Such a pattern did not change in the dark, but the nucleotide levels were proportionally higher than those observed in the light. Both wake-sleep and light-dark related changes were statistically significant [77]. Thus, it appears that cAMP concentration oscillates in the PO-AH under the influence of either a circadian or an ultradian factor, which act independently.

The cAMP changes during the modification of the wake-sleep cycle induced by exposure to Ta 0°C are shown in Fig. 5. This situation is similar to that observed in Fig. 3 since, in order to avoid conflicts arising from the possible influence of ambient factors, the changes in ambient temperature and the start of the experimental sessions were made coincident with the start of the light-dark cycle. Animals were sacrificed in the first hours of the light period following 48 h either at Ta 0°C or at normal laboratory Ta. Thus, in this experiment the exposure was a

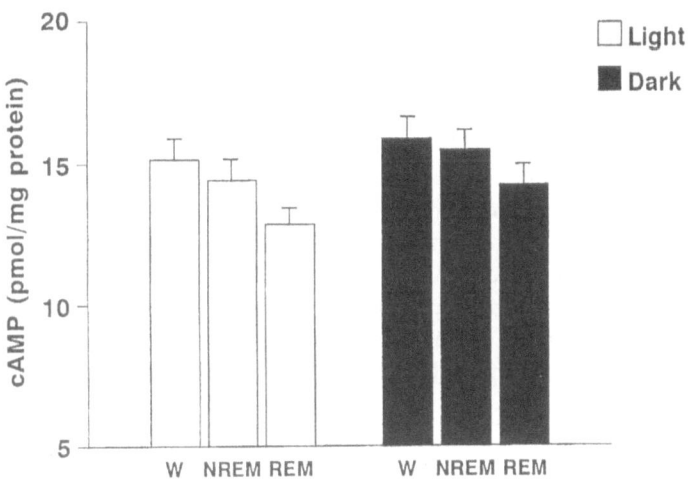

Fig. 4. *Average cAMP concentration in the preoptic-anterior hypothalamic area during the different stages of the wake-sleep cycle during either the light or dark period of the light-dark cycle.* Rats were kept under normal laboratory conditions. *Thin bar:* S.E.M. Statistical significance for W/NREM/REM differences, P<0.01; for light-dark differences, P<0.05. *W:* wakefulness; *NREM, REM:* respective sleep stages. Data calculated from the results of [77]

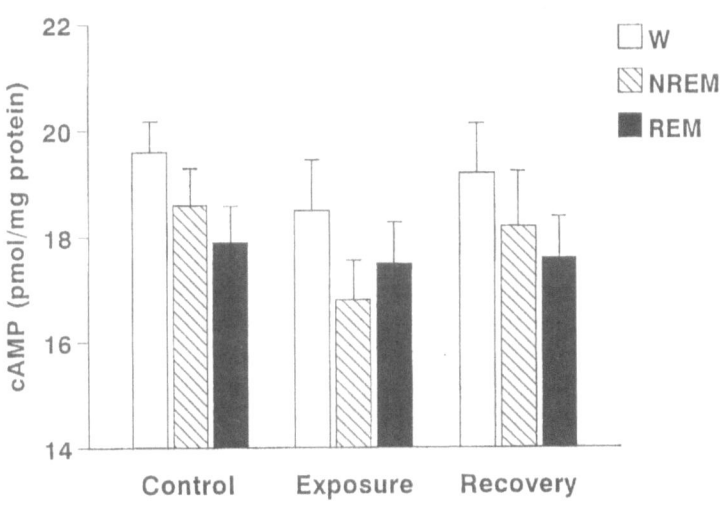

Fig. 5. *Average cAMP concentrations in the preoptic-anterior hypothalamic area of the rat during the different stages of the wake-sleep cycle in different experimental conditions. Control:* 49-53 h at normal ambient temperature (Ta 23°C); *Exposure:* 49-53 h at Ta 0°C; *Recovery:* 1-5 h at Ta 23°C following 48 h exposure to Ta 0°C. *Thin bar:* S.E.M. Statistical significance for W, NREM and REM differences, P<0.05. *W:* wakefulness; *NREM, REM:* respective sleep stages. Data calculated from the results of [78]

few hours longer than that depicted in Fig. 3, whilst the recovery consisted of only the first few hours of the light period. The results show that the ultradian oscillation in the PO-AH cAMP concentration, observed at normal laboratory Ta (control), is lost during the exposure to low Ta and reattained during the recovery. It is interesting to note that such changes concern only wakefulness and NREM sleep, whilst the levels of cAMP observed during REM sleep remained constant across all the experimental conditions. However, whilst the wake-sleep related changes in nucleotide levels were significant, those related to the ambient conditions were not [78]. On a statistical ground, this result may be easily explained by the fact that the changes in cAMP concentration induced by the exposure were masked by the successive increase during the recovery and, on a behavioural ground, by the fact that the wake-sleep cycle is much less disturbed during the second day of exposure to Ta 0°C (Fig. 3).

When the cAMP concentration in the PO-AH was measured in conditions having more intense influences on wake-sleep processes (Ta -10°C for 48 h), the changes observed were more clear-cut (Fig. 6). Since at such a low ambient temperature the wake-sleep cycle is characterized by few REM sleep episodes, that is, by a negligible number of complete cycles, the nucleotide levels were determined in wakefulness. At normal laboratory temperature there was a clear circadian rhythm of cAMP concentration, with the lowest values occurring during the light hours and the highest values occurring during the dark hours. For the first few hours, the exposure to Ta -10 °C did not affect either the nucleotide concentration nor its rhythmic change. However, the situation changed during the first dark period at low Ta: following an early rise, similar to that observed at normal laboratory Ta, the cAMP concentration declined to a low level, which was kept for the rest of the experimental condition. Such a decrease was concomitant with the disappearance of any significant circadian fluctuation in the nucleotide levels [79, 80].

Considered together with the results presented in Fig. 5, these data suggest that a disturbance in the occurrence of complete wake-sleep cycles, such as that encountered at low ambient temperatures, may be related to an impairment of the ultradian fluctuation of cAMP concentration at the PO-AH level. It also appears that at Ta -10°C such a reduction becomes progressively stronger since, following a few hours at such a low Ta, the decrease in the cAMP concentration is concomitant with the abolition of the circadian oscillation. Such a problem is addressed in a different way in Fig. 7. The diagrams show how the cAMP concentration, observed in the PO-AH during wakefulness, changes relatively, at the end of the exposure to either Ta 0°C or Ta -10°C and at the beginning of the respective recovery period, with respect to the values determined at normal laboratory temperature. The duration of the experimental periods was the same as that in Fig. 5. The decrease in cAMP concentration was relatively larger following the exposure to Ta -10°C than to Ta 0°C, and due to a relatively similar increase, the control nucleotide levels were reattained during the recovery following the exposure to Ta 0°C, but not during that following the exposure to Ta -10°C. Thus, as suggested by the course of REM sleep occurrence during the 48 h exposure to either Ta 0°C or Ta -10°C and the respective recovery periods (Fig. 3), the decrease in the PO-AH cAMP concentration appears

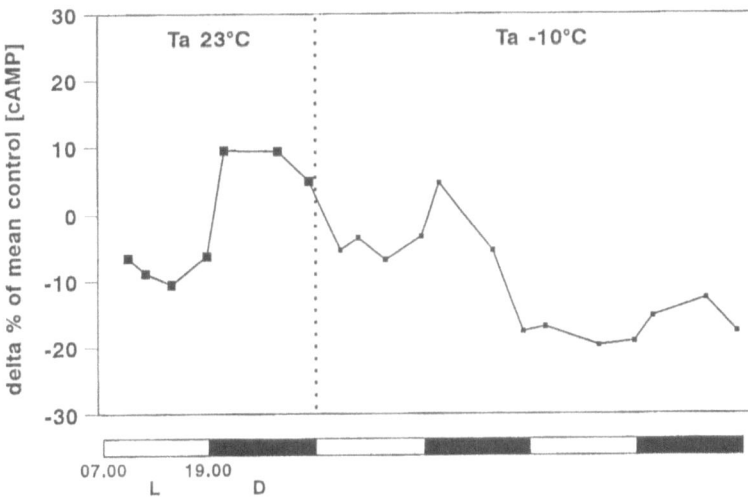

Fig. 6. *Average relative percent cAMP concentrations in the preoptic-anterior hypothalamic area of the rat.* cAMP levels were determined during wakefulness at different hours of the day, within either a 24 h period under normal laboratory conditions (*large filled squares*) or a 48 h of exposure to Ta -10°C (*small filled squares*). Data calculated from the results of [79, 80]

related to the degree of wake-sleep disturbance at low ambient temperature, whilst the amplitude of the differential of the concentration values, with respect to the control levels, appears to affect the organization of the wake-sleep cycle during the recovery. Although the situation depicted in Fig. 7 is clearly more static than that related to an ultradian determination of cAMP concentration, the cAMP concentration measurements, taken at the end of the exposure and at the beginning of the recovery, are close enough in time to suggest that the large decrease and the small successive increase found at Ta -10°C are signs of a dampening in the periodic variation in nucleotide accumulation.

Such an investigation has been carried out by testing, in the same experimental conditions as those depicted in the lower diagram of Fig. 7, the maximal capacity for the accumulation of cAMP at the PO-AH level [81] (Fig. 8). The cAMP accumulating system was maximally stimulated by exposing animals to acute hypoxia, a stimulus known to increase brain cAMP concentration in small rodents [81-83]. The histograms indicate the effects of the hypoxic stimulation in relative units of the cAMP concentration observed at normal laboratory temperature in normoxia. At the end of the exposure to Ta -10°C, the cAMP accumulating system was clamped at an activity level much lower than that in the control. Moreover, it appears that during first hours of the recovery period the cAMP accumulating system has not yet been restored to its full capacity. Thus, it appears that the occurrence of a normal wake-sleep cycle is somehow related to a normal capacity to accumulate cAMP at the PO-AH level. When this capacity is reduced, the normal wake-sleep cycle is impaired. Moreover, intermediate levels of PO-AH cAMP accumulation were concomitant with more particular disturbances. For

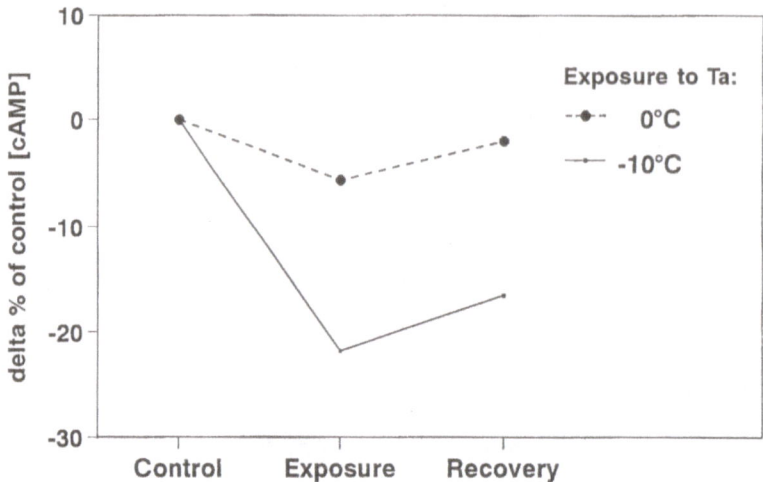

Fig. 7. *Average relative percent cAMP concentrations in the preoptic-anterior hypothalamic area of the rat.* cAMP levels were measured during wakefulness under different experimental conditions. *Control:* 49-53 h at normal ambient temperature (Ta 23°C); *Exposure:* 49-53 h at either Ta 0°C or Ta -10°C; *Recovery:* 1-5 h at Ta 23°C following 48 h of exposure to either Ta 0°C or Ta -10°C. Data calculated from the results of [78, 81]

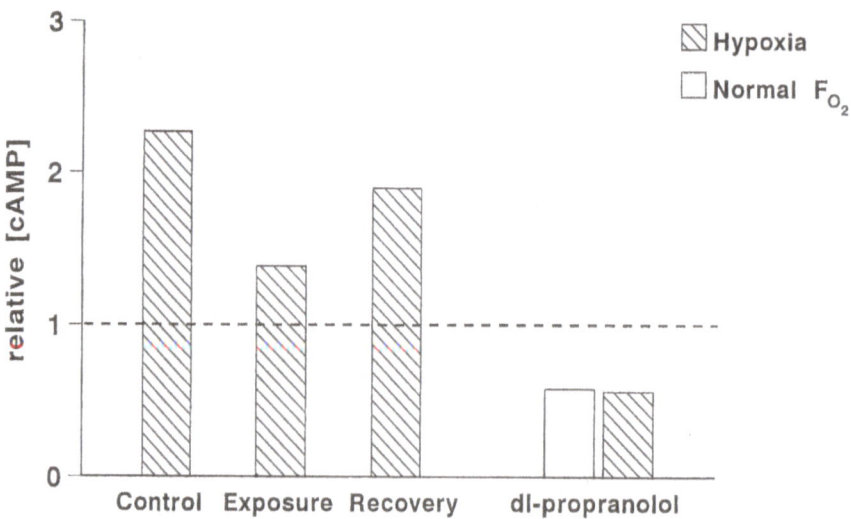

Fig. 8. *Average relative cAMP concentrations in the preoptic-anterior hypothalamic area of the rat.* cAMP levels were measured following 75 s of acute hypoxia in different experimental conditions. *Control:* 53 h at normal ambient temperature (Ta 23°C); *Exposure:* 53 h at Ta -10°C; *Recovery:* 5 h at Ta 23°C following 48 h of exposure to Ta -10°C. The short-term effects of the intraperitoneal administration of *dl*-propranolol (23 mg/kg) on cAMP accumulation, in either hypoxic or normoxic condition, are also shown. All data are referred to the basal normoxic cAMP levels (1, *dashed line*) which were determined in animals kept at Ta 23°C. F_{O_2}, fractional oxygen concentration. Data calculated from the results of [81, 83]

example, the recovery following a 48 h exposure to Ta -10°C was characterized by a normal amount of single REM sleep episodes, but by a low amount of sequential REM sleep episodes.

Figure 8 also shows the results of a further test of the importance of a normal level of cAMP accumulation in the PO-AH for wake-sleep processes. This experiment was carried out by administering *dl*-propranolol, a non-specific ß-adrenergic blocker, to the animals. Propranolol decreased the cAMP concentration in PO-AH to a level lower than that in the control, and also prevented the rise in the nucleotide accumulation caused by the hypoxic stimulation. Thus, it appears that *dl*-propranolol decreases cAMP concentration, since the nucleotide accumulation is clamped at a very low level in its presence. Examination of the wake-sleep cycle during the period of maximal efficacy of the drug action on cAMP accumulation showed that REM sleep was completely suppressed, NREM sleep reduced and wakefulness increased [83]. The role of post-synaptic mechanisms in the determination of the low cAMP accumulation capacity, observed during the exposure to Ta -10°C, is supported by the finding that the kinetic parameters (Km and $Vmax$) of tyrosine hydroxylase, the rate limiting enzyme in catecholamine synthesis, did not change in PO-AH following a two-day exposure to such a low Ta [84].

Conclusions

The changes in cAMP concentration observed in the preoptic-anterior hypothalamic area indicate that a certain level of cellular activity in this region is necessary to maintain a normal wake-sleep cycle. However, neither is PO-AH a sleep centre nor is cAMP a sleep factor. Although the changes in cAMP concentration in the PO-AH do not distinguish the effects of thermal stress from those of sleep deprivation, it may be hypothesized that both the response to an ambient challenge and the accumulation of a sleep debt are related to the same stressing condition. On this ground, changes in cAMP accumulation at the PO-AH level represent corresponding changes in cellular activity related to both ambient and sleep influences.

The evaluation of cellular activity, determined from changes in cAMP concentration, offers the possibility to follow nervous events which are in the time domain of the ultradian duration of the different stages of the wake-sleep cycle. Other longer, state-dependent events related to waking and sleeping may be monitored by changes in the transcription of the immediate-early genes which can be induced in minutes by extracellular stimuli [85]. For example, the c-*fos* proto-oncogene, particularly studied in the nervous system, changes in both mRNA and protein expression in relation to wake-sleep processes [86-88]. Such an approach has recently allowed detailed maps of cellular activity to be drawn in relation to sleep [89], and points to the importance of same region we have investigated. This is particularly interesting, since it is thought that c-*fos* induction may monitor intracellular second messenger levels [85].

References

1. Borbély AA, Tobler I (1989) Endogenous sleep-promoting substances and sleep regulation. Physiol Rev 69:605-670
2. Jones BE (1994) Basic mechanisms of sleep-wake states. In: Kryger MH, Roth T, Dement WE (eds) Principles and practice of sleep medicine. WB Saunders, London, pp 145-162
3. Kapás L, Obál FJr, Krueger JM (1993) Humoral regulation of sleep. Int Rev Neurobiol 35:131-160
4. Parmeggiani PL (1982) Regulation of physiological functions during sleep in mammals. Experientia 38:1405-1408
5. Parmeggiani PL (1994) The autonomic nervous system in sleep. In: Kryger MH, Roth T, Dement WE (eds) Principles and practice of sleep medicine. WB Saunders, London, pp 194-203
6. Parmeggiani PL, Morrison AR (1990) Alterations in autonomic functions during sleep. In: Loewy AD, Spyer KM (eds) Central regulation of autonomic functions. Oxford University Press, Oxford, pp 367-386
7. Moruzzi G (1972) The sleep-waking cycle. Rev Physiol 64:1-165
8. Loewy AD (1990) Anatomy of the autonomic nervous system: An overview. In: Loewy AD, Spyer KM (eds) Central regulation of autonomic functions. Oxford University Press, Oxford, pp 3-16
9. Loewy AD (1991) Forebrain nuclei involved in autonomic control. In: Holstege G (ed) Role of the forebrain in sensation and behavior. Prog Brain Res 87:253-268
10. Guyenet PC (1991) Central noradrenergic neurons: the autonomic connection. In: Barnes CD, Pompeiano O (eds) Neurobiology of the locus coeruleus. Prog Brain Res 88:365-380
11. Nieuwenhuys R (1985) Chemoarchitecture of the brain. Springer-Verlag, Berlin Heidelberg, pp 177-193
12. Nieuwenhuys R, Veening JG, Domburg PV (1988/89) Core and paracores: some new chemoarchitectural entities in the mammalian neuraxis. Acta Morphol Neerl Scand 26:131-163
13. Swanson LW (1986) Organization of mammalian neuroendocrine system. In: Mountcastle VB, Bloom FE, Geiger SR (eds) Intrinsic regulatory systems of the brain.(Handbook of physiology, sect 1, The nervous system, vol IV) American Physiological Society, Bethesda, pp 317-363
14. Swanson LW (1987) The hypothalamus. In: Björklund A, Hökfelt T, Swanson LW (eds) Integrated systems of the CNS. (Handbook of chemical neuroanatomy, part 1, vol 5) Elsevier, Amsterdam, pp 1-124
15. Butcher LL, Woolf NJ (1984) Histochemical distribution of acetylcholinesterase in the central nervous system: clues to the localization of cholinergic neurons. In: Björklund A, Hökfelt T, Kuhar MJ (eds) Classical transmitters and transmitter receptors in the CNS. (Handbook of chemical neuroanatomy, part II, vol 3). Elsevier, Amsterdam, pp 1-50
16. Björklund A, Lindvall O (1986) Catecholaminergic brain stem regulatory systems. In: Mountcastle VB, Bloom FE, Geiger SR (eds) Intrinsic regulatory systems of the brain (Handbook of physiology, sect 1, The nervous system, vol IV). American Physiological Society, Bethesda, pp 155-235
17. Striker EM, Zigmond MJ (1986) Brain monoamines, homeostasis and adaptive behavior. In: Mountcastle VB, Bloom FE, Geiger SR (eds) Intrinsic regulatory systems of the brain (Handbook of physiology, sect 1, The nervous system, vol IV). American Physiological Society, Bethesda, pp 677-700

18. Aghajanian GK, Vandermaelen CP (1986) Specific systems of the reticular core: Serotonin. In: Mountcastle VB, Bloom FE, Geiger SR (eds) Intrinsic regulatory systems of the brain (Handbook of physiology, sect 1, The nervous system, vol IV). American Physiological Society, Bethesda, pp 237-256

19. Hobson JA, Steriade M (1986) Neuronal basis of behavioral state control. In: Mountcastle VB, Bloom FE, Geiger SR (eds) Intrinsic regulatory systems of the brain (Handbook of physiology, sect 1, The nervous system, vol IV). American Physiological Society, Bethesda, pp 701-823

20. Giuditta A (1977) The biochemistry of sleep. In: Davison AN (ed) Biochemical correlates of brain structure and function. Academic Press, New York, pp 293-337

21. Karnovsky ML, Reich P (1977) Biochemistry of sleep. In: Agranoff BW, Aprison MH (eds) Advances in neurochemistry, vol 2. Plenum, New York, pp 213-275

22. Giuditta A, Perrone Capano C, Grassi Zucconi G (1984) The neurochemical approach to the study of sleep. In: Lajtha A (ed) Handbook of neurochemistry, vol 8. Plenum, New York, pp 443-476

23. Madsen PL, Vorstrup S (1991) Cerebral blood flow and metabolism during sleep. Cerebrovasc Brain Metab Rev 3:281-296

24. Franzini C (1992) Brain metabolism and blood flow during sleep. J Sleep Res 1: 3-16

25. Sokoloff L (1996) Cerebral metabolism and visualization of cerebral activity. In: Greger R, Windhorst V (eds) Comprehensive human physiology, vol 1. Springer-Verlag, Berlin Heidelberg, pp 579-602

26. Fox PT, Raichle ME, Mintum MA, Deuce C (1988) Nonoxidative glucose consumption during focal physiologic neural activity. Science 241:462-464

27. Mata M, Fink DJ, Gainer H, Smith CB, Davidsen L, Savaki H, Schwartz WJ, Sokoloff L (1980) Activity-dependent energy metabolism in rat posterior pituitary primarily reflects sodium pump activity. J Neurochem 34:214-215

28. Kadekaro M, Crane AM, Sokoloff L (1985) Differential effects of electrical stimulation of sciatic nerve on metabolic activity in spinal cord and dorsal root ganglion in the rat. Proc Natl Acad Sci USA 82:6010-6013

29. Ramm P, Frost BJ (1983) Regional metabolic activity in the rat brain during sleep-wake activity. Sleep 6:196-216

30. Ramm P, Frost BJ (1986) Cerebral and local cerebral metabolism in the cat during slow wave and rem sleep. Brain Res 365:112-124

31. Abrams RM, Hutchinson AA, Jay TM, Sokoloff L, Kennedy C (1988) Local cerebral glucose utilization non-selectively elevated in rapid eye movement sleep of the fetus. Dev Brain Res 40:65-70

32. Maquet P, Dive D, Salmon E, Sadzot B, Franco G, Poirrier R, von Frenckell R, Franck G (1990) Cerebral glucose utilization during sleep-wake cycle in man determined by positron emission tomography and [^{18}f]2-fluoro-2-deoxy-D-glucose method. Brain Res 513:136-143

33. Lydic R, Baghdoyan HA, Hibbard L, Bonyak EV, DeJoseph MR, Hawkins RA (1991) Regional brain glucose metabolism is altered during rapid eye movement sleep in the cat: a preliminary study. J Comp Neurol 304:517-529

34. Siggins GR, Gruol DL (1986) Mechanisms of transmitter action in the vertebrate central nervous system. In: Mountcastle VB, Bloom FE, Geiger SR (eds) Intrinsic regulatory systems of the brain (Handbook of physiology, sect 1, The nervous system, vol IV). American Physiological Society, Bethesda, pp 1-114

35. Hemmings HC Jr, Nairn AC, McGuinnes TL, Huganir RL, Greengard P (1989) Role of protein phosphorylation in neuronal signal transduction. FASEB J 3:1583-1592

36. Parmeggiani PL (1980) Temperature regulation during sleep: A study in homeostasis.

In: Orem J, Barnes CD (eds) Physiology in sleep. Academic Press, London, pp 97-143

37. Parmeggiani PL (1988) Thermoregulation during sleep from the viewpoint of home-ostasis. In: Lydic R, Biebuyek JF (eds) Clinical physiology of sleep. American Physiological Society, Bethesda, pp 159-169

38. Parmeggiani PL, Rabini C (1967) Shivering and panting during sleep. Brain Res 6:789-791

39. Zepelin H (1994) Mammalian sleep. In: Kryger MH, Roth T, Dement WE (eds) Principles and practice of sleep medicine. WB Saunders, London, pp 69-80

40. Boulant JA (1980) Hypothalamic control of thermoregulation. In: Morgane PJ, Panksepp J (eds) Behavioral studies of the hypothalamus.(Handbook of the hypo-thalamus, vol 3, part A) Marcel Dekker, New York Basel, pp 1-82

41. Simon E, Pierau F-K, Taylor DCM (1986) Central and peripheral thermal control of effectors in homeothermic temperature regulation. Physiol Rev 66:235-300

42   Crosby EC, Showers MJC (1969) Comparative anatomy of the preoptic and hypothal-amic areas. In: Haymaker W, Anderson E, Nauta WJH (eds) The hypothalamus. Charles C Thomas, Springfield, Illinois, pp 61-135

43. Prosser CL (ed) (1991) Comparative animal physiology: Neural and intergrative ani-mal physiology. Wyley-Liss, New York

44. Johnson KG, Hales JRS (1984) An introductory analysis of competition between ther-moregulation and other homeostatic systems. In: Hales JRS (ed) Thermal physiology. Raven, New York, pp 295-298

45. Parmeggiani PL, Rabini C (1970) Sleep and environmental temperature. Arch Ital Biol 108:369-387

46. Parmeggiani PL, Rabini C, Cattalani M (1969) Sleep phases at low environmental temperature. Arch Sci Biol 53:277-290

47. Glotzbach SF, Heller HC (1994) Temperature regulation. In: Kryger MH, Roth T, Dement WE (eds) Principles and practice of sleep medicine. WB Saunders, London, pp 260-275

48. Benington JH, Heller HC (1994) REM-sleep timing is controlled homeostatically by accumulation of REM-sleep propensity in non-REM sleep. Am J Physiol 266:R1992-R2000

49. Borbély AA (1994) Sleep homeostasis and models of sleep regulation. In: Kryger MH, Roth T, Dement WE (eds) Principles and practice of sleep medicine. WB Saunders, London, pp 309-320

50. Rosenberg RS, Bergmann BM, Rechtschaffen AA (1976) Variations in slow wave activ-ity during sleep in the rat. Physiol Behav 17:931-938

51. Trachsel L, Tobler I, Borbély AA (1988) Electroencephalogram analysis of non-rapid eye movement sleep in rats. Am J Physiol 255:R27-R37

52. Borbély AA, Neuhaus HV (1979) Sleep-deprivation: effects on sleep and EEG in the rat. J Comp Physiol 133:71-87

53. Friedman L, Bergmann BM, Rechtschaffen AA (1979) Effects of sleep deprivation on sleepiness, sleep intensity, and subsequent sleep in the rat. Sleep 1:369-391

54. Borbély AA, Tobler I, Hanagasioglu M (1984) Effect of sleep deprivation on sleep and EEG power spectra in the rat. Behav Brain Res 14:171-182

55. Franken P, Dijk D-J, Tobler I, Borbély AA (1991) Sleep deprivation in rats: effects on EEG power spectra, vigilance states and cortical temperature. Am J Physiol 261:R198-R208

56. Amici R, Zamboni G, Perez E, Jones CA, Toni I, Culin F, Parmeggiani PL (1994) Pattern of desynchronized sleep during deprivation and recovery induced in the rat by changes in ambient temperature. J Sleep Res 3:250-256

57. Trachsel L, Tobler I, Acherman P, Borbély A (1991) Sleep continuity and the REM-nREM cycle in the rat under baseline conditions and after sleep deprivation. Physiol Behav 49:575-580
58. Kripke DF, Reite ML, Pegram LM, Stephens LM, Lewis OF (1968) Nocturnal sleep in rhesus monkeys. Electroencephalogr Clin Neurophysiol 24:582-586
59. Ursin R (1970) Sleep stage relations within the sleep cycles of the cat. Brain Res 20:91-97
60. Kobayashi T, Tsuji Y, Endo S (1985) Sleep cycles as a basic unit of sleep. In: Schultz H, Lavie P (eds) Ultradian rhythms in physiology and behavior. Exp Brain Res suppl 12. Springer-Verlag, Berlin Heidelberg, pp 260-269
61. Merica H, Gaillard JM (1991) A study of the interrupted REM episode. Physiol Behav 50:1153-1159
62. Amici R, Zamboni G, Perez E, Jones CA, Parmeggiani PL (1997) The influence of a heavy thermal load on REM sleep in the rat. Brain Res (submitted)
63. Sakaguchi S, Gltozbach SF, Heller HC (1979) Influence of hypothalamic and ambient temperatures on sleep in kangaroo rats. Am J Physiol 237:R80-R88
64. Roussel B, Turrillot P, Kitahama K (1984) Effect of ambient temperature on sleep-waking cycle in two strains of mice. Brain Res 294:67-73
65. Sichieri R, Schmidek WR (1984) Influence of ambient temperature on the sleep-wakefulness cycle in the golden hamster. Physiol Behav 33:871-877
66. Parmeggiani PL, Cianci T, Calasso M, Zamboni G, Perez E (1980) Quantitative analysis of short term deprivation and recovery of desynchronized sleep in cats. Electroencephalogr Clin Neurophysiol 50:293-302
67. Siegel GJ, Agranoff BW, Albers RW, Molinoff PB (1994) Basic neurochemistry. Raven, New York
68. Cooper DMF, Mons N, Karpen JW (1995) Adenylyl cyclases and the interreaction between calcium and cAMP signalling. Nature 374:421-424
69. Mons M, Cooper D (1995) Adenylate cyclases: critical foci in neuronal signaling. Trends Neurosci 18:536-542
70. Sutherland EW (1972) Studies on the mechanism of hormone action. Science 177:401-408
71. Beavo JA (1995) Cyclic nucleotide phosphodiesterases functional implications of multiple isoforms. Physiol Rev 75:725-748
72. Francis SH, Corbin JD (1994) Structure and function of cyclic nucleotide-dependent protein kinases. Annu Rev Physiol 56:237-272
73. Zimmerman AL (1995) Cyclic nucleotide gated channels. Curr Opin Neurobiol 5:296-303
74. Hunter T (1995) Protein kinases and phosphatases: the yin and yang of protein phosphorylation and signaling. Cell 80:225-236
75. Jessel TM, Kandel ER (1993) Synaptic transmission: a bidirectional and self-modifiable form of cell-cell communication. Cell 72/ Neuron 10[Suppl]:1-30
76. Steriade M (1994) Brain electrical activity and sensory processing during waking and sleep states. In: Kryger MH, Roth T, Dement WE (eds) Principles and practice of sleep medicine. WB Saunders, London, pp 105-124
77. Perez E, Zamboni G, Amici R, Fadiga L, Parmeggiani PL (1991) Ultradian and circadian changes in the cAMP concentration in the preoptic region of the rat. Brain Res 551:132-135
78. Perez E, Zamboni G, Amici R, Jones CA, Parmeggiani PL (1995) cAMP accumulation in the hypothalamus, cerebral cortex, pineal gland and brown fat across the wake-sleep cycle of the rat exposed to different ambient temperatures. Brain Res 684:56-60

79. Perez E, Zamboni G, Parmeggiani PL (1982) cAMP concentration in the rat's preoptic region and cerebral cortex during sleep deprivation and recovery induced by ambient temperature. Exp Brain Res 47:114-118

80. Zamboni G, Perez E, Parmeggiani PL (1982) Cyclic AMP concentration in the rat's preoptic region. Experientia 38:1188-1189

81. Zamboni G, Jones CA, Amici R, Perez E, Parmeggiani PL (1996) The capacity to accumulate cyclic AMP in the preoptic-anterior hypothalamic area of the rat is affected by the exposition to low ambient temperature and the subsequent recovery. Exp Brain Res 109:164-168

82. Gross RA, Ferrendelli JA (1980) Mechanisms of cyclic AMP regulation in cerebral anoxia and their relationship to glycogenolysis. J Neurochem 34:1309-1318

83. Zamboni G, Perez E, Amici R, Parmeggiani PL (1990) The short-term effects of dl-propranolol on the wake-sleep cycle of the rat are related to selective changes in preoptic cyclic AMP concentration. Exp Brain Res 81:107-112

84. Perez E, Amici R, Bacchelli B, Zamboni G, Libert JP, Parmeggiani PL (1987) Kinetic parameters of tyrosine hydroxylase activity during sleep deprivation and recovery induced by ambient temperature. Sleep 10:436-442

85. Morgan JI, Curran T (1991) Stimulus-transcription coupling in the nervous system: involvement of the inducible proto-oncogenes fos and jun. Annu Rev Neurosci 14:421-451

86. Pompeiano M, Cirelli C, Tononi G (1992) Effects of sleep deprivation on Fos-like immunoreactivity in the rat brain. Arch Ital Biol 130:325-335

87. Pompeiano M, Cirelli C, Tononi G (1994) Immediate-early genes in spontaneous wakefulness and sleep: expression of c-fos and NGFI-A mRNA and protein. J Sleep Res 3:80-96

88. O'Hara BF, Young KA, Watson FL, Heller HC, Kilduff TS (1993) Immediate early gene expression in brain during sleep deprivation: preliminary observations. Sleep 16:1-7

89. Sherin JE, Shiromani PJ, McCarley RW, Saper CB (1996) Activation of ventrolateral preoptic neurons during sleep. Science 271:216-221

Regulation of Cerebral and Extracerebral Circulation During the Wake-Sleep Cycle

P. Lenzi, G. Zoccoli and C. Franzini

Introduction

A discontinuity characterizes the sleep process: the steady unfolding of quiet sleep (non-rapid eye movement sleep, NREMS) is periodically interrupted by active sleep episodes (rapid eye movement sleep, REMS). The state change from NREMS to REMS has best been described as switching from a homeostatic to a non-homeostatic condition [1]; in the latter the physiological control loops are affected by disturbances (neural discharges in sensory systems and efferent pathways) of central origin locked to the sleep process. Not all the control loops are equally affected, and "central commands" and set point values may also be perturbed; at any rate, a condition of physiological instability appears to be the cardinal feature of REMS. This "physiological risk" [2] may be related to the high incidence of acute cardiovascular disease in the early morning hours of higher REMS density [3-5]. The substantial overlap of neural circuits in the brain stem responsible for controlling both cardiovascular regulation and REMS phasic events gives a "structural" foundation to cardiovascular instability during REMS [6] (Fig. 1).

Control Mechanisms of the Cerebral Circulation During the Wake-Sleep Cycle

The control of cerebral circulation basically relies on the local coupling between neuronal activity and blood flow (*flow-metabolism couple*). Moreover, cerebral blood flow (CBF) is maintained adequate to cerebral metabolic needs even in the presence of changes in perfusion pressure (*autoregulation*), and is strongly affected by changes in blood gases (*chemical regulation*). The main function of CBF is the regulation of tissue homeostasis, i.e. to maintain "optimal" tissue concentrations of solutes that are relevant for brain metabolism, in the different conditions of cerebral activity. Solute concentrations depend on the balance of (a) local metabolic solute production and consumption, and (b) blood solute supply

Dipartimento di Fisiologia Umana e Generale, Università di Bologna, Piazza di Porta S. Donato 2, 40127 Bologna, Italy

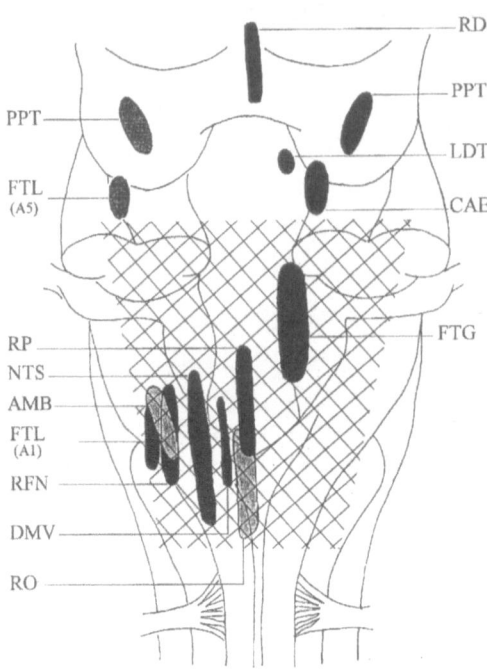

Fig. 1. *Schematic representation of cat brain stem showing neuronal groups involved in cardiovascular regulation (left) and in the control of different components of REM sleep phenomenology (right).* In the two sets some nuclei coincide; others belong to a commonly identified anatomical circuitry. Cross-hatching corresponds to the pressor and depressor centers [92]. *AMB:* n. ambiguus; *CAE:* n. caeruleus; *DMV:* dorsal motor n. of the vagus; *FTG:* gigantocellular tegmental field; *FTL:* lateral tegmental field (two noradrenergic neuronal groups, A1 and A5, are indicated); *LDT:* laterodorsal tegmental n.; *NTS:* n. tractus solitarii; *PPT:* n. tegmenti peduncolo-pontinus (Kölliker-Fuse n.); *RD:* n. raphe dorsalis; *RFN:* retrofacial n. (rostral ventrolateral medulla); *RO:* n. raphe obscurus; *RP:* n. raphe pallidus. Nomenclature from [93]

and its removal which result from blood-brain barrier permeability, blood solute concentrations and blood flow to the tissue. Blood flow, in turn, depends on perfusion pressure and local vascular resistance. Many changes in concentrations of relevant solutes can occur, but only one effector mechanism, the regulation of vascular resistance, maintains brain tissue homeostasis in physiological conditions; only one degree of freedom is available for correction. As a consequence, despite optimization of the vascular effector, it is impossible to maintain all solute concentrations at their optimal level. Major errors will be partially corrected, but at the expense of producing many new, minor errors. The results of all cerebrovascular adjustments occurring in response to changes in local metabolic rate, blood solute composition and perfusion pressure can be explained on a single basis. While aiming at optimal tissue concentrations of solutes, the system must strike a balance between correcting the original error and generating new errors. Substances produced by brain metabolism (e.g. CO_2 and H^+) are an example. If systemic blood concentration of a substance is increased by extracerebral mechanisms, this will determine (a) a parallel increase in tissue concentration of the same substance, which in turn will increase blood flow, and (b) as a consequence of the increase in blood flow, an "undesired" change in tissue concentrations of other solutes. Similar considerations hold when nutrient (e.g. O_2, glucose) concentrations in blood are reduced. These "undesirable side effects" should be kept in mind when searching for substances responsible for blood flow modifications. The causal relationship works in two ways: a concentration change of a given metabolite can be the cause or effect of blood flow change.

From the data obtained to date in different conditions of wake (sensory stimulation [7], mental activation [8], quiet wakefulness[9]) and sleep (NREMS and REMS [10]), similar changes occur in blood flow, oxygen consumption and glucose uptake; moreover, no basic differences emerge in the mechanisms linking CBF to brain metabolism. Thus, results obtained in different wake-sleep states can be tentatively pooled to draw an overall picture of flow-metabolism coupling in the brain. The brain normally extracts about 50% of oxygen and 10% of glucose from the blood (for a review see [11]). During functional activation, glucose uptake and CBF undergo parallel, marked increases (51% and 50%, respectively [7]), while O_2 uptake increases slightly (5%) and tissue oxygen concentration rises above control values, since CBF increase exceeds the increase in O_2 uptake. Extracellular glucose concentration undergoes an early decrease (~10-30%), followed by an increase of about the same magnitude [12]. Comparable changes are observed in the transition from NREMS to REMS: glucose uptake and CBF increase, while the arteriovenous difference in O_2 content is reduced [10, 13].

The above data show that O_2 is inadequate as a mediator between blood flow and metabolism, since tissue O_2 concentration is increased during activation (both in wake and REMS states) and its arteriovenous difference is reduced in REMS. Glucose concentration, on the other hand, could provide such a link: glucose uptake and blood flow in fact undergo parallel increases (51% and 50% [7], 12% and 14% [8]), the vascular response developing within 1 s from the beginning of brain activation [14]. However, the low glucose extraction coefficient weighs against this possibility. On the contrary, extracellular hydrogen ions, adenosine and other metabolites cannot be ruled out. It is unlikely that the link between blood flow and metabolic activity is restricted to a single constituent of the extracellular fluid; different substances, alone or in combination, will drive flow according to the specific biochemical needs of the brain in each physiological or pathophysiological condition.

Only limited experimental data exist on systemic factors (PaO_2, $PaCO_2$, blood pressure) which might affect the control of cerebral circulation during sleep:

a) In NREMS a slight hypercapnia develops, which counteracts the vascular effects of the decreased cerebral metabolic rate, and accounts for the modest increase in CBF in some species (goat [15]). $PaCO_2$ becomes an important determinant of CBF regulation during sleep in pathologic conditions (sleep apnea syndrome [16, 17]).

b) In REMS the tonic increase in CBF with respect to NREMS values is independent of blood pressure (BP) changes: a comparative study across species shows that the CBF surge occurs in the face of decrements, increments or no changes in BP [13]. Moreover, the range of variations of BP during REMS is well within the limits of autoregulation which has been shown to operate in all sleep states [18]. REMS, however, is characterized by a high metabolic rate, high CBF, and consequently reduced vasodilatory reserve which may place the brain at risk for ischemic hypoxia during acute hypotension, partly accounting for cardiovascular morbidity and mortality during sleep.

The independence of CBF from systemic hemodynamics is further supported by the lack of correlation between blood flow changes in the brain and in other peripheral circulations (kidney, muscle, skin, splanchnic): CBF is not affected by the redistribution of regional flows occurring in REMS [19]. Further, an independent regulation of CBF and extracerebral carotid circulation has been suggested during this sleep state in rats [20]. Taken together, these data confirm a local regulation of CBF during REMS, and local metabolic factors remain the most probable candidates accounting for the CBF surge in REMS.

In conclusion, the efficacy of both autoregulation and flow-metabolism coupling results in the brain circulation being less subjected to the instabilities during REMS which affect other functions (respiration, thermoregulation) and other regions of the peripheral vascular bed.

Sleep-Dependent Changes in Cerebral Circulation

The different levels of vigilance are associated with different levels of neuronal activity. Overall, neuronal firing decreases slightly from wakefulness to NREMS, and increases markedly from NREMS to REMS: metabolic and CBF changes follow accordingly [10, 13] (Fig. 2). Flowmeter recordings show that the CBF increase in REMS comprises both tonic and phasic components [21]. This is a general feature of blood flow changes in REMS; the fast surges are temporally related to other central or peripheral phasic events (ponto-geniculo-occipital waves, rapid eye movements, muscle twitches). Studies of regional CBF evidence higher increments in brainstem and diencephalic structures compared to the cerebral cortex and the cerebellum [13]. Since brain stem and diencephalic networks are responsible for the occurrence of REMS episodes, this confirms the importance of the *neuronal activity/blood flow* coupling mechanism in driving CBF changes during REMS.

Finally, the density of perfused capillaries in the brain is maximal and stable across vigilance states in normal [22] (Fig. 3), hypertensive [23] and aged [23a] rats. Changes in CBF differentially affect different transports across the blood-brain barrier. In fact, a CBF increment always favors flow-limited transport; but when the CBF increment is accompanied by a capillary surface increase (*capillary recruitment:* opening of previously unperfused capillaries), diffusion-limited transport is also enhanced since diffusion depends on permeability and surface area. No change in the number of perfused capillaries occurs in the transition

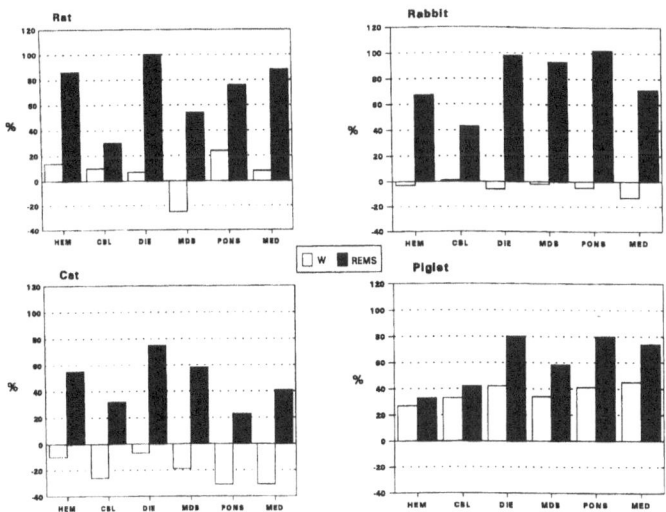

Fig. 2. *Regional cerebral blood flows during wakefulness (W) and REM sleep (REMS) are expressed as percent changes from NREM sleep values.* During *REMS cerebral blood flow increases significantly in all regions and species considered. HEM: cerebral hemispheres; CBL: cerebellum; DIE: diencephalon; MDB: midbrain; MED: medulla.* Data from [20, 19, 94, 78], respectively

from NREMS to REMS: the CBF increase in REMS should favor, therefore, the flow-limited transcapillary transfer of substances such as oxygen and carbon dioxide.

Pathophysiological Implications

REMS is a condition of high CBF and low vasodilatory reserve confronting the systemic challenges of BP instabilities and PaO_2 and $PaCO_2$ shifts. It is worth mentioning here that these challenges increase during obstructive sleep apnea episodes (cf. [24]), and the risk of cerebrovascular disease increases accordingly [25].

Control Mechanisms of the Extracerebral Circulation During the Wake-Sleep Cycle

The study of mutual interactions between sleep and the autonomic nervous system has a long and distinguished tradition in sleep physiology, since it bears on the understanding of both the phenomenology [26] and the functions [27] of sleep and has been extensively reviewed [28-30]. In recent years the widespread application of two techniques, namely spectral analysis of heart rate (HR) and BP variability in humans and other species and sympathetic nerve activity recordings in

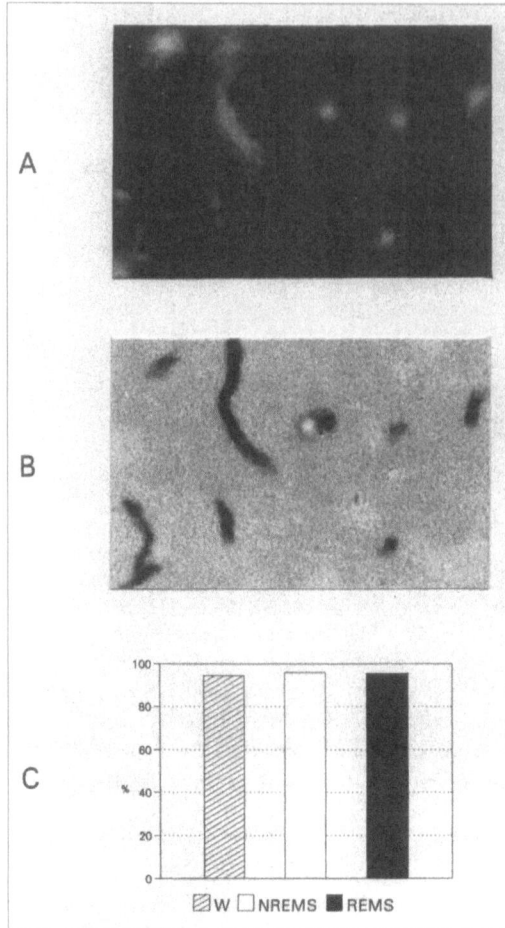

Fig. 3a-c. *Brain capillary perfusion during the wake-sleep cycle in rat.* a. Perfused capillaries labeled by intravascular Evans Blue. b. Same field after alkaline phosphatase stain of anatomical capillary. c. Average percent values of anatomical capillaries which are perfused in the different vigilance states: *W:* wakefulness; *NREMS:* non-REM sleep; *REMS:* REM sleep. It can be seen that capillary perfusion is not affected by vigilance states. Data from [22]

humans, prompted a series of new studies. The results of these investigations and their methodological caveats are addressed below.

Spectral Analysis of Heart Rate and Blood Pressure Variability

An important step in evaluation of cardiovascular control mechanisms came from the recognition that oscillations in HR and BP result both from the feedback operation of the regulatory loops and from central commands; the frequency content of these fluctuations can be assessed by spectral analysis [31, 32] (for a critical review of the topic cf. [33]). HR and BP power spectra, however, depict the modulation of autonomic control on sinoatrial node and vascular smooth muscle, not necessarily its absolute value. In many physiological conditions modulation amplitude is proportional to absolute value [34, 35]. However, experimental manipulations have shown that this does not always apply: during hypertension induced by phenylephrine [36], the absolute value of low frequency spectral power did not change with respect to control conditions.

In HR power spectra, the low frequency band (LF, < 0.15 Hz) has mainly been associated with the modulation of sympathetic outflow, while the high frequency band (HF, > 0.15 Hz) has been referred to the modulation of parasympathetic output [31, 32, 35, 37]. The vast majority of spectral analysis studies during sleep refer to HR spectra; in this case, the inference of overall autonomic changes during sleep implies the tacit and unwarranted assumption of parallel modifications in cardiac and vasomotor sympathetic outflow. BP power spectra are affected by the modulation of both cardiac output and peripheral resistances and therefore they can provide only a spurious index of the modulation of peripheral vasomotion: a contribution of parasympathetic efferent activity to the LF band has been recognized [32, 38]. Moreover, the concept of a uniform "sympathetic tone" has been challenged [39], and regional differences occur during sleep [40].

Studies of HR power spectra during sleep [41-48] report reasonably concordant results (Fig. 4): during NREMS the LF component decreases and the HF component increases with respect to wakefulness (W) levels. During REMS both the LF and HF components return toward the W levels; overshoots and undershoots have been described, to maximal values of the LF components [42, 46] and to zero value of the HF components [42, 47], respectively. The contribution of sympathetic and parasympathetic efferent activity to LF and HF power spectra, respectively, has been assessed during sleep by experiments of selective pharma-

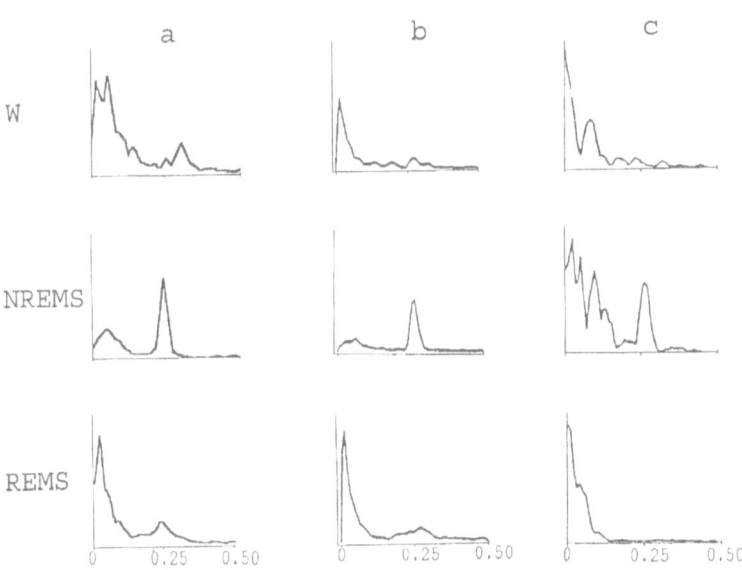

Fig. 4a-c. *Spectral analysis during the wake-sleep cycle.* Power spectra from different studies are shown. In the same vigilance state, qualitative similitudes and differences in power spectra from different studies can be appreciated. Ordinates are in arbitrary units, abscissae in Hz. *W:* wakefulness; *NREMS:* non-REM sleep; *REMS:* REM sleep. **a.** Modified from [50]. **b.** Modified from [46]. **c.** Modified from [47]

cological blockade (propranolol, atropine) [47]. No changes in the LF [45] or HF [46] components in the transition from NREMS to REMS have also been reported, together with species-related [49] or age-related [50] exceptions to the above pattern of power peaks. A different analysis of HR variability (Poincaré plots) confirmed an increased vagal activity during NREMS [51].

Only a few studies have assessed BP power spectra during sleep. In man [44] a decrease of the LF component parallels an increase of the HF component during NREMS with respect to wakefulness. In REMS only the HF component returns toward the waking levels, while the LF component remains low; this contrasts with the increment of the LF component of heart rate variability, and suggests the possibility of dissociation between cardiac and vasomotor sympathetic activity during REMS. In cat, a prominence of the LF component during REMS has been described [52, 53].

Continuous 24 h monitoring of HR and BP in man evaluates the autonomic markers of cardiovascular control in normal life, outside the laboratory setting. These studies, however, do not allow a precise polygraphic assessment of sleep stages, and therefore are of limited value in clarifying specific sleep-related physiological or pathophysiological mechanisms. On the whole, during the night power spectra related to sympathetic control decrease, together with the increase in power spectra related to parasympathetic activity [54, 55]. However, more complex circadian patterns have been described [56], warning against a simplistic, reductive interpretation of power spectra changes.

Sympathetic Nerve Activity Recordings

Multi-unit recordings of sympathetic nerve vasomotor activity during sleep were obtained both in cats and in humans. In renal nerves of normal [57, 58] and decerebrate [40, 59] cat, and in splanchnic nerves of decerebrate cat [40] sympathetic vasomotor activity decreased during REMS; muscle nerve sympathetic vasomotor activity increased during REMS in decerebrate cat [40]. In humans, muscle nerve vasoconstrictor activity decreased during NREMS and increased during REMS [60-63]. Cutaneous vasodilator activity was unchanged from wakefulness to NREMS, but increased in REMS [64].

The above data weigh in favor of the differentiation of sympathetic efferent activity, and against the concept of a global sympathetic tone, during REMS (cf. [39, 65]). Moreover, given the multiple hierarchic controls on the autonomic effector, a strict correspondence in efferent-effector activity is often lacking [39]. This is well exemplified in the control of muscle circulation during REMS: blood flow may increase, decrease or remain unchanged [66, 67] in the face of a stereotyped increase in muscle nerve sympathetic activity.

Sleep-dependent Changes in Extracerebral Circulation

The data hereafter presented derive from experiments on different animal species (sheep, pig, dog, cat, rabbit, rat) in which blood flow was measured with flowmeters or with radioactive microspheres. No data are available on sleep-state dependent changes in peripheral extracerebral circulation in humans.

Muscle Circulation

Muscle blood flow does not change significantly in the transition from wakefulness to NREMS. During REMS sympathetic vasomotor activity to muscle blood vessels increases [40, 68]. Moreover, REMS is characterized by specific changes in muscle activity: atonia and twitches. The flow-activity couple therefore entails the opposite effects of vasoconstriction (atonia) and vasodilation (twitches). The cumulative effect of these factors determines blood flow changes in the two muscle fiber populations:

a) The slow (red) fibers have a high basal blood flow in NREMS: further vasodilator influence (twitching activity) is without effect, while vasoconstrictor influence (sympathetic vasomotion, atonia) reduces the vessel caliber. The net effect is a reduced blood flow in REMS.

b) The fast (white) fibers have a low basal flow in NREMS; slight quantitative differences in vasodilator (local) and vasoconstrictor (neural and local) mechanisms may result in increased (rabbit), decreased (rat) or unchanged (cat) blood flow in REMS [66, 67, 69].

Therefore, sleep-dependent shifts in the balance of neural and local factors underlie parallel or opposite muscle blood flow changes during REMS in the two muscle fiber populations.

Splanchnic Circulation

No significant blood flow changes occur in the splanchnic territory during sleep, either in cat [70] or in rabbit [71]. The sleep process however interferes with the complex interplay of local and neural factors controlling regional circulation. Local factors (metabolites and products of digestion) are likely to be sleep-independent, whereas a decrease in mesenteric nerve vasomotor activity occurs during REMS [40]. The basal value of sympathetic nerve activity in NREMS, and hence the extent of its drop in REMS, may vary in different species; accordingly, species differences in vascular conductance (increased in cat [70], unchanged in rabbit [71]) have been reported.

In liver circulation, the mixture of arterial and portal blood flow is precisely regulated: an increase in hepatic arterial flow is compensated by a decrease in portal flow and vice versa, thus tending to a constancy of total liver blood flow. The reciprocity of the two constituents of hepatic blood flow entails a significant negative correlation between arterial and portal components during wakefulness and NREMS; the correlation disappears in REMS [71]. This can be viewed as a fine example of a specific regulatory mechanism disrupted by the "disturbances" (breathing irregularities, changes in abdominal pressure, cardiovascular variability) arising in REMS.

Cutaneous Circulation

During NREMS a sleep-dependent decrease in set-point temperature from values in wakefulness underlies the regulated blood flow increase in cutaneous circulation in cat [72], rabbit [73], rat [74] and in humans [64]. In REMS, on the contrary, skin blood flow changes reveal the impairment of the thermoregulatory effector response [75]: in a cold environment a blood flow increment results from the drop in vasoconstrictor tone; in a warm environment a blood flow reduction results from a decrease in BP (cat [72]) or a decrease in neurogenic vasodilation (rabbit [73], rat [74]). Since the regulation of cutaneous blood flow depends mainly on neural vasomotor activity, the above data confirm a general paradigm: the sleep-dependent blood flow change in the different vascular beds correlates with the relative weight of their neural control.

Renal Circulation

Very few studies exist on renal circulation during sleep. Kidney blood flow is not state-dependent in cat [76, 77] or pig [78]. When, however, a thermal load imposes renal vasoconstriction as part of an integrated thermoregulatory vasomotor pattern, this neural vasomotion, like any other thermoregulatory effector response [75], subsides in REMS, and blood flow increases [79].

Coronary Circulation

In dog, left coronary blood flow decreases significantly from wakefulness to NREMS and increases in REMS [80]. In particular:
a) During phasic REMS blood flow surges are coupled with episodes of sinus tachycardia, and coronary vascular resistance decreases concurrently [81]; both tachycardia and blood flow surges are eliminated by bilateral stellectomy. Changes in the magnitude of heart rate and blood flow increments are closely matched suggesting that metabolic vasoactive substances are responsible for reduced coronary vascular resistance, with the sympathetic nervous system initiating the sequence: increased frequency-increased metabolic activity-increased flow.
b) In the transition from NREMS to REMS an increase in coronary blood flow follows a pause in heart rhythm [82]; the increase in flow without a corresponding increase in cardiac metabolic demand may indicate a neurogenically mediated vasomotion. Further studies are needed to clarify the role of α-adrenergic vasoconstrictor [81] and cholinergic vasodilator influences [82] contributing to coronary vasomotion during REMS.
   When, on the other hand, a marked stenosis is induced by cuff inflation in the left circumflex coronary artery, REMS surges in heart rate are accompanied by a

decrease in coronary blood flow. A reduced diastolic perfusion time may underlie the flow decrement [83]. Regional blood flow distribution during sleep in piglets [78] confirms that the increment in left ventricular blood flow during REMS is present at a very early age (6 days).

The above studies highlight the complexity of phasic and tonic sympathovagal interactions in determining cardiac excitability during REMS. Results of the few experiments of stimulation [84], lesion [80, 84] and recording [85, 86] of autonomic nerves to the heart are not easily reconciled in a unitary picture.

Integrated Vasomotor Patterns

The circulatory system is endowed with control mechanisms hierarchically organized in levels of increasing complexity (intrinsic-extrinsic; lower integrative-higher integrative). A priori, it can be surmised that the higher integrative levels of control are those most likely to be affected by the sleep process [30]. To test this hypothesis, a stimulus is needed (thermal load, hemorrhage) capable of driving the most complex level of the regulatory response (adaptive redistribution of regional flows).

Rabbits sleeping in a non-neutral environment [79] during NREMS show the same thermoregulatory vasomotor adjustments present in wakefulness. At low ambient temperature (Ta), vasoconstriction of the splanchnic and renal beds and the arteriovenous anastomoses (AVA) accompanies muscle vasodilation (coupled with shivering thermogenesis. At high Ta, respiratory muscle vasodilation (coupled with panting) and opening of AVAs entail compensatory vasoconstriction in the splanchnic and renal beds. This blood flow redistribution abates in REMS and, as a consequence, flow is reduced in vasodilated beds and increased in constricted beds: the compensatory modulation of peripheral resistances in response to a thermal load is lost (Fig. 5). Another vasomotor pattern, namely the gradient of muscle vasoconstriction, increasing from superficial to deeper layers and corresponding to the reduction of the inner thermoregulated "core" in a cold environment, is lost during REMS [87].

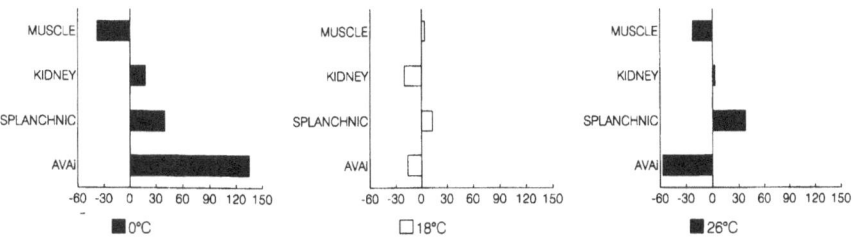

Fig. 5. *Regional blood flow (RBF) percent changes from NREM sleep (NREMS) to REM sleep (REMS) at low (left), neutral (center) and high (right) ambient temperature (Ta).* RBF changes when entering REMS are not stereotyped, since they are determined by the disappearance of the adaptive vasomotor pattern preexisting in NREMS and are different at differing Ta. *AVAi:* index of arteriovenous anastomotic flow. Modified from [79]

In lambs [88], following hemorrhage, cardiac output is uniformly reduced in the different vigilance states but during REMS the compensatory increase in peripheral vascular resistance fails, entailing a maximal BP drop (Fig. 6). This cannot result from baroreflex impairment, since the heart rate response following hemorrhage is greater during REMS than during NREMS or wakefulness.

The inadequate response of the peripheral circulation is sleep-dependent, and results from the impairment of hypothalamic integrative activity during REMS [30]: the hypothalamus is unresponsive to both thermal [89] and electrical [90] stimulation, and recordings of hypothalamic neurons show that cellular thermosensitivity is attenuated or abolished [91]. While we do not understand the meaning of the sudden changes in neuronal activity, metabolism and flow that accompany REMS [13], it is apparent that they are not compatible with the integrative function of hypothalamic structures; complex adaptive vasomotor patterns are affected as a consequence.

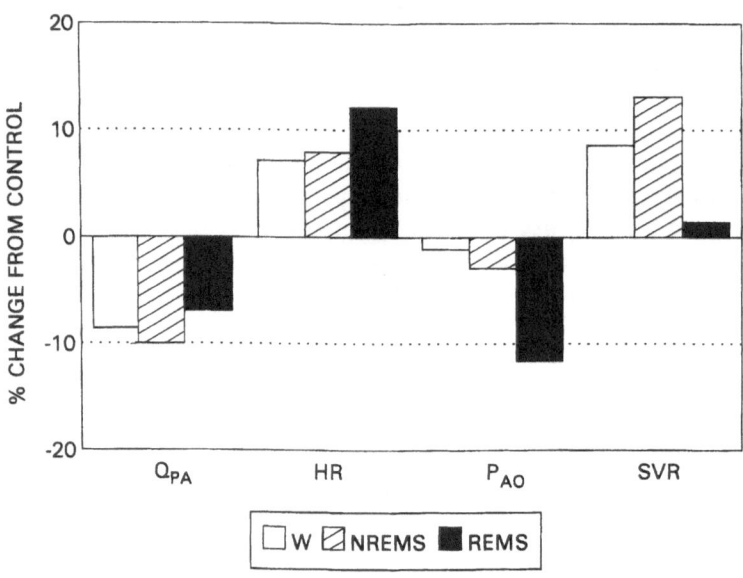

Fig. 6. *Percent change from control values of cardiovascular variables following 10 ml/kg hemorrhage during wakefulness (W), NREM sleep (NREMS) and REM sleep (REMS).* In REMS the hemorrhage is not compensated by an increased peripheral resistance, with a consequent fall in blood pressure. Q_{PA}: pulmonary artery blood flow; *HR*: heart rate; P_{AO}: aortic blood pressure; *SVR*: systemic vascular resistance. Modified from [88]

Pathophysiological Implications

An apparent contradiction results from the reviewed data: during REMS important modifications in autonomic control entail relatively modest changes at the effector level (myocardium, vascular smooth muscle); this reflects the redundancy of local and neural control of the circulation. However, the few studies considering integrated vasomotor patterns indicate that, when the autonomic response is essential to challenge an external (thermal [79]) or internal (volemic [88]) perturbation, the regulatory deficit stands out. It may be surmised that the loss of integrative control of peripheral circulation during REMS will precipitate pathological cardiovascular events when the redistribution of peripheral blood flow is a major compensating mechanism (e.g., after myocardial ischemia, or in chronic heart failure).

References

1. Parmeggiani PL (1980) Behavioral phenomenology of sleep (somatic and vegetative). Experientia 36:6-11
2. Parmeggiani PL (1991) Physiological risks during sleep. In: Peter JH, Penzel T, Podzus T, von Wichert P (eds) Sleep and health risk. Springer-Verlag, Berlin, Heidelberg, pp 119-123
3. Muller JE, Stone PH, Turi ZG, Rutherford JD, Czeisler CA, Parker C, Poole WK, Passmani E, Roberts R, Sobel BE, Willerson JT , Braunwald E (1985) Circadian variation in the frequency of onset of acute myocardial infarction. N Engl J Med 313:1315-1322
4. Muller JE, Ludmer PL, Willich SN, Tofler JH, Aylmer G, Klangos I , Stone PH (1987) Circadian variation in the frequency of sudden cardiac death. Circulation 75:131-138
5. Marler JR, Price TR, Clark GL, Muller JE, Robertson T, Mohr J, Hier DB, Wolf PA, Caplan MR , Foulkes MA (1989) Morning increase in onset of ischemic stroke. Stroke 20:473-476
6. Franzini C, Zoccoli G, Cianci T , Lenzi P (1996) Sleep-dependent changes in regional circulations. News Physiol Sci 11:274-280
7. Fox PT, Raichle ME, Mintun MA, Dence C (1988) Nonoxidative glucose consumption during focal physiologic neural activity. Science 241:462-464
8. Madsen PL, Hasselbach SG, Hageman LP, Skovgaard Olsen K, Bulow J, Holm S, Wildschiodtz G, Paulson OB, Lassen NA (1995) Persistent resetting of the cerebral oxygen/glucose uptake ratio by brain activation: evidence obtained with the Kety-Schmidt technique. J Cereb Blood Flow Metab 15:485-491
9. Madsen PL (1993) Blood flow and oxygen uptake in the human brain during various states of sleep and wakefulness. Acta Neurol Scand 88, Suppl. 148:1-27
10. Madsen PL, Vorstrup S (1991) Cerebral blood flow and metabolism during sleep. Cerebrovasc Brain Metab Rev 3:281-296
11. Villringer A, Dirnagl U (1995) Coupling of brain activity and cerebral blood flow: basis of functional neuroimaging. Cerebrovasc Brain Metab Rev 7:240-276
12. Silver I, Erecinska M (1994) Extracellular glucose concentration in mammalian brain: continuous monitoring of changes during increased neuronal activity and upon limitation in oxygen supply in normo-, hypo-, and hyper-glycemic animals. J Neurosci 14:5068-5076
13. Franzini C (1992) Brain metabolism and blood flow during sleep. J Sleep Res 1:3-16
14. Frostig RD, Lieke EE, Ts'o DY, Grinvald A (1990) Cortical functional architecture and local coupling between neuronal activity and the microcirculation revealed by in

vivo high-resolution optical imaging of intrinsic segnals. Proc Natl Acad Sci USA 87:6082-6086

15. Santiago TV, Guerra E, Neubauer JA, Edelman NH (1984) Correlation between ventilation and brain blood flow during sleep. J Clin Invest 73:497-506

16. Hajak G, Klingelhofer J, Schulz-Varszegi M, Staedt J, Conrad B, Ruther E (1995) Cerebral perfusion during sleep-disordered breathing. J Sleep Res 4, Suppl 1:117-124

17. Siebler M, Nachtmann A (1993) Cerebral hemodynamics in obstructive sleep apnea. Chest 103:1118-1119

18. Walker A, Grant D, Wild J, Franzini C (1994) Cerebral autoregulation during sleep in lambs. J Sleep Res 3, Suppl. 1:273

19. Lenzi P, Cianci T, Guidalotti PL, Leonardi GS, Franzini C (1987) Brain circulation during sleep and its relation to extracerebral hemodynamics. Brain Res 415:14-20

20. Zoccoli G, Bach V, Cianci T, Lenzi P, Franzini C (1994) Brain blood flow and extracerebral carotid circulation during sleep in rat. Brain Res 641:46-50

21. Grant DG, Franzini C, Wild J, Walker AM (1995) Continuous measurement of blood flow in the superior sagittal sinus of the lamb. Am J Physiol 269:R274-R279

22. Zoccoli G, Lucchi ML, Andreoli E, Bach V, Cianci T, Lenzi P, Franzini C (1996) Brain capillary perfusion during sleep. J Cereb Blood Flow Metab 16:1312-1318

23. Zoccoli G, Lucchi ML, Andreoli E, Cianci T, Lenzi P, Franzini C (1996) Brain capillary perfusion during sleep in the spontaneously hypertensive rat. J Sleep Res 5, Suppl. 1:514

23a. Zoccoli G, Lucchi ML, Andreoli E, Lenzi P, Franzini C. Brain capillary perfusion during sleep in aged rat. (In preparation)

24. Shepard JW (1994) Cardiorespiratory changes in obstructive sleep apnea. In: Kryger MH, Roth T, Dement WC (eds) Principles and practice of sleep medicine. WB Saunders, Philadelphia, pp 657-666

25. Palomaki H, Partinen M, Erkinjuntti T, Kaste M (1992) Snoring, sleep apnea syndrome and stroke. Neurology 42 (suppl.6):75-82

26. Kleitman N (1963) Sleep and wakefulness. The University of Chicago Press, Chicago, pp 1-552

27. Hess WR (1949) Das Zwischenhirn - Syndrome, Lokalisationen, Functionen. Benno Schwabe, pp 1-187

28. Parmeggiani PL (1984) Autonomic nervous system in sleep. Exp Brain Res 54 [Suppl. 8]:39-49

29. Parmeggiani PL, Morrison AR (1990) Alterations in autonomic functions during sleep. In: Loewy AD, Spyer KM (eds) Central regulation of autonomic functions. Oxford University Press, Oxford, pp 367-386

30. Parmeggiani PL (1994) The autonomic nervous system in sleep. In: Kryger MH, Roth T, Dement WC (eds) Principles and practice of sleep medicine. WB Saunders, London, pp 194-203

31. Akselrod S, Gordon D, Ubel FA, Shannon DC, Barer AC, Cohen RJ (1981) Power spectrum analysis of heart rate fluctuation: a quantitative probe of beat-to-beat cardiovascular control. Science 213:220-222

32. Akselrod S, Gordon D, Madwed JB, Snidam NC, Shannon DC, Cohen RT (1985) Hemodynamic regulation: investigation by spectral analysis. Am J Physiol 249:H867-H875

33. Parati G, Saul JP, Di Rienzo M, Mancia G (1995) Spectral analysis of blood pressure and heart rate variability in evaluating cardiovascular regulation. A critical appraisal. Hypertension 25:1276-1286

34. Pagani M, Lombardi G, Guzzetti S, Rimoldi O, Furlan R, Pizzinelli P, Sandrone G,

Malfatto G, DellOrto S, Piccaluga E, Turiel M, Baselli G, Cerutti S, Malliani A (1986) Power spectral analysis of heart rate and arterial pressure variabilities as a marker of sympatho-vagal interaction in man and conscious dog. Circ Res 59:178-193

35. Pomeranz B, Macaulay RJB, Shannon DC, Cohen RJ, Benson H (1985) Assessment of autonomic functions in humans by heart rate spectral analysis. Am J Physiol 248:H151- H153

36. Saul JP, Rea RF, Eckberg DL, Berger RD, Cohen RJ (1990) Heart rate and muscle sympathetic nerve variability during reflex changes of autonomic activity. Am J Physiol 258:H713-H721

37. Malliani A, Pagani M, Lombardi F, Cerutti S (1991) Cardiovascular neural regulation explored in the frequency domain. Circulation 84:482-492

38. Saul JP, Berger RD, Albrecht P, Stein P, Chen MH, Cohen RJ (1991) Transfer function analysis of the circulation: unique insights into cardiovascular regulation. Am J Physiol 261:H1231-H1245

39. Wallin GB, Elam M (1994) Insights from intraneural recordings of sympathetic nerve traffic in humans. News Physiol Sci 9:203-207

40. Futuro-Neto HA, Coote JH (1982) Changes in sympathetic activity to heart and blood vessels during desynchronized sleep. Brain Res 252:259-268

41. Baharav A, Kotagal S, Gibbons V, Rubin BK, Pratt G, Karin J, Akselrod S (1995) Fluctuations in autonomic nervous activity during sleep displayed by power spectrum analysis of heart rate variability. Neurology 45:1183-1187

42. Berlad I, Shiltner A, Ben-Haim S, Lavie P (1993) Power spectrum analysis and heart rate variability in stage 4 and REM sleep: evidence for state-specific changes in autonomic dominance. J Sleep Res 2:88-90

43. Negoescu RM, Csiki IE (1989) Autonomic control of the heart in some vagal maneuvers and normal sleep. Rev Roum Morphol Embryol Physiol Physiologie 26:39-49

44. Van de Borne P, Nguyen H, Biston P, Linkowski P, Degaute JP (1994) Effects of wake and sleep stages on the 24-h autonomic control of blood pressure and heart rate in recumbent men. Am J Physiol 266:H548-H554

45. Vanoli E, Adamson PB, Ba-Lin MBH, Pinna GD, Lazzara R, Orr WC (1995) Heart rate variability during specific sleep stages. A comparison of healthy subjects with patients after myocardial infarction. Circulation 91:1918-1922

46. Vaughn BV, Quint SR, Messenheimer JA, Robertson KR (1995) Heart period variability in sleep. Electroencephalogr Clin Neurophysiol 94:155-162

47. Zemaityte D, Varoneckas G, Sokolov E (1984) Heart rhythm control during sleep. Psychophysiology 21:279-289

48. Zemaityte D, Varoneckas G, Plauska K, Kaukenas J (1986) Components of the heart rhythm power spectrum in wakefulness and individual sleep stages. Int J Psychophysiol 4:129-141

49. Haddad GG, Jeng HJ, Lee SH, Lai TL (1984) Rhythmic variations in R-R interval during sleep and wakefulness in puppies and dogs. Am J Physiol 247:H67-H73

50. Finley JP, Nugent ST (1995) Heart rate variability in infants, children and young adults. J Autonom Nerv Syst 51:103-108

51. Raetz SC, Richard CA, Garfinkel A, Harper RM (1992) Dynamic characteristics of cardiac R-R intervals during sleep and waking states. Sleep 14:526-533

52. Kanamori N, Sakai K, Sei H, Salvert D, Vanni-Mercier G, Yamamoto M, Jouvet M (1994) Power spectral analysis of blood pressure fluctuations during sleep in normal and decerebrate cat. Arch Ital Biol 132:105-115

53. Kanamori N, Sakai K, Sei H, Bouvard A, Salvert D, Vanni-Mercier G, Jouvet M (1995) Effects of decerebration on blood pressure during paradoxical sleep in cats. Brain Res Bull 37:545-549

54. Di Rienzo M, Castiglioni P, Mancia G, Parati P, Pedotti A (1989) 24h sequential spectral analysis of arterial blood pressure and pulse interval in free-moving subjects. IEEE Trans Biomed Eng 36:1066-1075

55. Furlan R, Guzzetti S, Crivellaro W, Dassi S, Tinelli M, Baselli G, Cerutti S, Lombardi F, Pagani M, Malliani A (1990) Continuous 24-hour assessment of the neural regulation of systemic arterial pressure and RR variabilities in ambulant subjects. Circulation 81:537-547

56. Parati G, Castiglioni P, Di Rienzo M, Omboni S, Pedotti A, Mancia G (1990) Sequential spectral analysis of 24-hour blood pressure and pulse interval in humans. Hypertension 16:414-421

57. Baust W, Weidinger H, Kirchner F (1968) Sympathetic activity during natural sleep and arousal. Arch Ital Biol 106:379-390

58. Reiner PD (1986) Correlational analysis of central neuronal activity and sympathetic tone in behaving cats. Brain Res 378:86-96

59. Iwamura Y, Uchino Y, Ozawa S, Torii S (1969) Spontaneous and reflex discharge of a sympathetic nerve during "para-sleep" in decerebrate cat. Brain Res 16:359-367

60. Hornyak M, Cejnar M, Elam M, Matousek M, Wallin BG (1991) Sympathetic muscle nerve activity during sleep in man. Brain 114:1281-1295

61. Okada H, Iwase S, Mano T, Sugiyama Y, Watanabe T (1991) Changes in muscle sympathetic nerve activity during sleep in humans. Neurology 41:1961-1966

62. Shimizu T, Takahashi Y, Suzuki K, Kogawa S, Tashino T, Takahashi K, Hishikawa Y (1992) Muscle nerve sympathetic activity during sleep and its change with arousal response. J Sleep Res 1:178-185

63. Somers VK, Dyken ME, Mark AL, Abboud FM (1993) Sympathetic-nerve activity during sleep in normal subjects. N Engl J Med 328:303-307

64. Noll G, Elam M, Kunimoto M, Karlsson T, Wallin BG (1994) Skin sympathetic nerve activity and effector function during sleep in humans. Acta Physiol Scand 151:319-329

65. Mancia G (1993) Autonomic modulation of the cardiovascular system during sleep. N Engl J Med 328:347-349

66. Reis DJ, Moorhead D, Wooten GF (1969) Differential regulation of blood flow to red and white muscle in sleep and defense behavior. Am J Physiol 217:541-546

67. Lenzi P, Cianci T, Leonardi GS, Martinelli A, Franzini C (1989) Muscle blood flow changes during sleep as a function of fibre type composition. Exp Brain Res 74:549-554

68. Baccelli G, Albertini R, Mancia G, Zanchetti A (1974) Central and reflex regulation of sympathetic vasoconstrictor activity to limb muscles during desynchronized sleep in the cat. Circ Res 35:625-635

69. Zoccoli G, Lalatta Costerbosa G, Bach V, Andreoli E, Cianci T, Lenzi P, Franzini C (1994) Muscle blood flow during the sleep-wake cycle in the rat. J Sleep Res 3, Suppl. 1:283

70. Mancia G, Adams DB, Baccelli G, Zanchetti A (1969) Regional blood flows during desynchronized sleep in the cat. Experientia 25:48-49

71. Cianci T, Zoccoli G, Lenzi P, Franzini C (1990) Regional splanchnic blood flow during sleep in the rabbit. Pfluegers Arch 415:594-597

72. Parmeggiani PL, Zamboni G, Cianci T, Calasso M (1977) Absence of thermoregulatory vasomotor responses during fast wave sleep in cats. Electroencephalogr Clin Neurophysiol 42:372-380

73. Franzini C, Lenzi P, Cianci T, Guidalotti PL (1982) Neural control of vasomotion in rabbit ear is impaired during desynchronized sleep. Am J Physiol 243:R142-R146

74. Alfoldi P, Rubicsek G, Cserni G, Obal F (1990) Brain and core temperatures and peripheral vasomotion during sleep and wakefulness at various ambient temperatures in the rat. Pfluegers Arch 417:336-341

75. Parmeggiani PL (1980) Temperature regulation during sleep: a study in homeostasis. In: Orem J, Barnes CD (eds) Physiology in sleep. Academic Press, New York, pp 97-143

76. Mancia G, Baccelli G, Adams DB, Zanchetti A (1971) Vasomotor regulation during sleep in the cat. Am J Physiol 220:1086-1093

77. Mancia G, Baccelli G, Zanchetti A (1974) Regulation of renal circulation during behavioral changes in the cat. Am J Physiol 227:536-542

78. Cote A, Haddad GG (1990) Effect of sleep on regional blood flow distribution in piglets. Pediatr Res 28:218-222

79. Cianci T, Zoccoli G, Lenzi P, Franzini C (1991) Loss of integrative control of peripheral circulation during desynchronized sleep. Am J Physiol 261:R373-R377

80. Kirby DA, Verrier RL (1989) Differential effects of sleep stage on coronary hemodynamic function. Am J Physiol 256:H1378-H1383

81. Dickerson LW, Huang HA, Thurnher MM, Nearing BD, Verrier RL (1993) Relationship between coronary hemodynamic changes and the phasic events of REM sleep. Sleep 16:550-557

82. Dickerson LW, Huang AH, Nearing BD, Verrier RL (1993) Primary coronary vasodilation associated with pauses in heart rhythm during sleep. Am J Physiol 264:R186-R196

83. Kirby DA, Verrier RL (1989) Differential effects of sleep stage on coronary hemodynamic function during stenosis. Physiol Behav 45:1017-1020

84. Baust W, Bohnert B (1969) The regulation of heart rate during sleep. Exp Brain Res 7:169-180

85. Leichnetz GR (1972) Relationship of spontaneous vagal activity to wakefulness and sleep in the cat. Exp Neurol 35:194-210

86. Varbanova A, Nikolov N, Doneshka P (1974) Fluctuations in the vagal and sympathetic tone connected with the circadian cycle and the sleep-wakefulness cycle. Agressologie 16:23-33

87. Zoccoli G, Cianci T, Lenzi P, Franzini C (1992) Shivering during sleep: relationship between muscle blood flow and fiber type composition. Experientia 48:228-230

88. Fewell JE, Williams BJ, Hill DE (1984) Behavioral state influences the cardiovascular response to hemorrhage in lambs. J Dev Physiol 6:339-348

89. Parmeggiani PL, Franzini C, Lenzi P, Zamboni G (1973) Threshold of respiratory responses to preoptic heating during sleep in free moving cats. Brain Res 52:189-201

90. Parmeggiani PL, Calasso M, Cianci T (1980) Respiratory effects of preoptic-anterior hypothalamic electrical stimulation during sleep in cats. Sleep 4:71-82

91. Parmeggiani PL, Cevolani D, Azzaroni A, Ferrari G (1987) Thermosensitivity of anterior hypothalamic-preoptic neurons during the waking-sleeping cycle: A study in brain functional states. Brain Res 415:79-89

92. Alexander RS (1946) Tonic and reflex functions of medullary sympathetic cardiovascular centers. J Neurophysiol 9:205-217

93. Berman AL (1968) The brain stem of the cat. Univ. Wisconsin Press, pp1-175

94. Reivich M, Isaacs G, Evarts E, Kety S (1968) The effect of slow wave sleep and REM sleep on regional cerebral blood flow in cats. J Neurochem 15:301-306

Effects of Systemic Circulation on Hypothalamic Temperature in the Behavioural States of Sleep

P. L. PARMEGGIANI AND A. AZZARONI

Introduction

Hypothalamic temperature changes may be quantitatively expressed as the ratio between the changes in heat content and the mass of the hypothalamus multiplied by the specific heat of the tissue ($\Delta T = \Delta Q/Mc$). Heat is produced by cellular metabolism and is transferred to the perfusing arterial blood which is always maintained at a lower temperature than the hypothalamic tissue [1]. Hypothalamic temperature is constant when hypothalamic heat content is unchanged as a result of a perfect balance between heat production and heat loss.

In several mammalian species hypothalamic temperature undergoes small regular changes (a few tenths of a degree) in relation to the stages of sleep [1], i.e. it decreases during quiet sleep (QS; synonyms: synchronized sleep, NREM sleep) with respect to quiet wakefulness (QW), and increases during active sleep (AS; synonyms: desynchronized sleep, REM sleep) with respect to QS regardless of the exposure to a wide range of ambient temperatures [2-6] above and below ambient thermal neutrality [7].

The metabolic heat production of the hypothalamic tissue and the flow and/or the temperature of the arterial blood perfusing the hypothalamus are the main factors underlying these temperature changes. However, changes in hypothalamic metabolic rate would not be expected to contribute much to temperature variation since energy metabolism is coupled to heat clearance by blood flow (blood flow-metabolism coupling) [1, 8]. Therefore, changes in arterial blood flow and/or temperature, i.e. the mechanisms for brain cooling, become the fundamental points at issue [1, 9]. This chapter addresses the experimental evidence that shows the systemic circulatory events in the behavioural states of the ultradian sleep cycle affecting the temperature of the hypothalamus regardless of ambient thermal loads. This is a major physiological issue, since this brain region (and the adjoining preoptic region) has specific thermosensitive neurons directly influencing thermoregulatory mechanisms.

Dipartimento di Fisiologia Umana e Generale, Università di Bologna, Piazza di Porta S. Donato 2, 40127 Bologna, Italy

Mechanisms for Brain Cooling

There are different mechanisms for cooling the brain in mammals and, in some species, more than one mechanism may be present (Fig. 1a). In, for example, cat, dog, sheep, and goat, the arterial blood perfusing the brain is cooled through the systemic venous blood return to the heart (systemic brain cooling), where the cool venous blood from the systemic heat exchangers of the body (upper airway mucosa, ear pinna, horn, tail) mixes completely with the warm venous blood returning from body tissues. In addition, as shown by Hayward and Baker [1], there is a mechanism of selective brain cooling in which the carotid blood supply to the brain is thermally conditioned prior to entering the circle of Willis. Countercurrent heat exchange is achieved by a network of fine vessels (the carotid rete), derived from a branch of the external carotid artery (mainly the internal maxillary artery), and interlaced with vessels of the pterygoid venous plexus receiving cool venous blood from the systemic heat exchangers of the head (ear pinna, nasal mucosa, horn) [1, 10]. The carotid rete is connected to the circle of

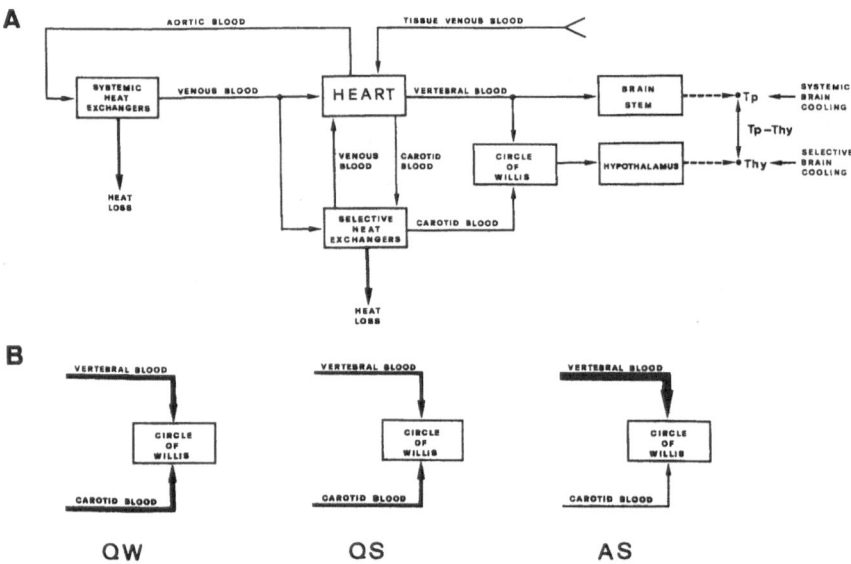

Fig. 1 a, b. *Mechanisms of systemic and selective brain cooling in the cat, a species which has a carotid rete.* **a.** Only the routes of blood flow thermally relating systemic (ear pinna, upper airway mucosa) and selective (carotid rete-venous plexus) heat exchangers with the heart and the encephalon are indicated. The cool venous blood returning to the heart from systemic heat exchangers mixes with warm venous blood returning from body tissues. Vertebral arterial blood is warmer than the carotid arterial blood entering the circle of Willis since the latter is additionally cooled by the selective heat exchangers. *Tp:* Temperature of the pons. *Thy:* Temperature of the hypothalamus. **b.** The carotid and vertebral arterial blood supply to the circle of Willis during quiet wakefulness (*QW*), quiet sleep (*QS*) and active sleep (*AS*) are shown schematically

Willis through a short anastomotic artery (homologous to the distal part of the internal carotid artery in species lacking the carotid rete). Within the carotid rete-venous plexus (i.e., a selective heat exchanger) [1, 10], heat is transferred from the warmer arterial blood (aortic arch temperature) to the cooler venous blood returning from heat-dissipating systemic heat exchangers of the head [1]. Therefore, the degree to which the temperature of carotid blood reaching the circle of Willis is decreased, with respect to that of the aortic arch blood, also depends on the degree of systemic heat exchanger vasodilation affecting the supply and loss of heat from the blood to the environment [1]. Conversely, vertebral artery blood is not thermally conditioned by a countercurrent heat exchange mechanism and enters the circle of Willis at the same temperature as the blood leaving the aortic arch [1]. The difference between vertebral blood temperature (systemic cooling only) and carotid blood temperature (both systemic and selective cooling) is determined primarily by the heat loss from the carotid rete and is, therefore, a quantitative indicator of the intensity of selective brain cooling (Fig. 1a) [11].

Another mechanism for selective brain cooling, typical of species lacking the carotid rete (e.g. the rabbit and the rat), is provided by the conductive heat exchange between the basal portion of the brain (including the circle of Willis and the hypothalamus) and the pterygoid plexus and the ophthalmic sinus which drain cool venous blood from the systemic heat exchangers of the head [12].

In the case of primates, however, the changes in heat loss from systemic heat exchangers, affecting carotid blood temperature only through the systemic venous return to the heart, are probably the most important determinants of hypothalamic temperature (systemic brain cooling). As yet, there is no consensus of opinion as to whether a mechanism for selective brain cooling plays a significant role in humans [13-16].

Factors Affecting Brain Cooling

Posture hydrostatically influences the vascular transmural pressure in systemic and selective heat exchangers. However, the actual caliber of blood vessels is determined not only by transmural pressure but also by the vasoconstrictor sympathetic outflow. Therefore, the influences of negative or positive hydrostatic loads (head-up above or head-down below the heart level) and of vasoconstrictor sympathetic outflow may interact agonistically or antagonistically to determine the eventual caliber of blood vessels. The degree of the resulting vasodilation or vasoconstriction is an important variable underlying the amount of heat dissipated by systemic and selective heat exchangers.

An increase in the tonic vasoconstrictor sympathetic outflow to systemic heat exchangers decreases both systemic and selective brain cooling [17]. The difference between vertebral and carotid blood temperatures, i.e. the effect of selective brain cooling, is reduced since the heat loss from the carotid rete to the venous plexus is decreased as a result of the higher temperature of the venous blood returning from the vasoconstricted systemic heat exchangers (Fig. 2). In theory,

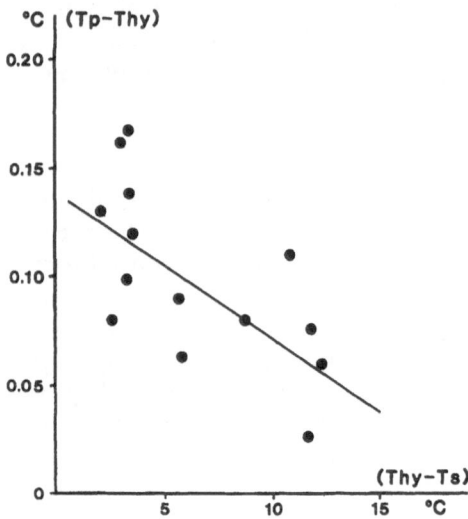

Fig. 2. *Changes in selective brain cooling as a function of systemic heat exchanger vasocon- striction during both quiet wakefulness and quiet sleep.* The intensity of selective brain cooling is appraised by the difference (*Tp-Thy*) between temperature of the pons (*Tp*) and temperature of the hypothalamus (*Thy*). In particular, *Thy* is affected by the temperature of the mixture of carotid blood and vertebral blood in the circle of Willis, whereas *Tp* is affected solely by the temperature of the vertebral blood. Therefore, under analogous metabolic heat production conditions, (*Tp-Thy*) is determined primarily by the tempera- ture and flow rate of carotid artery blood perfusing the circle of Willis. The difference (*Thy-Ts*) between *Thy* and ear pinna temperature (*Ts*) is a net indicator of the intensity of tonic vasoconstrictor sympathetic outflow to systemic heat exchangers which is necessary to maintain *Thy* homeostasis during quiet wakefulness and quiet sleep. (*Thy-Ts*) is mainly affected by changes in *Ts* which are of the order of degrees (°C) and dependent on local vasomotion. In comparison, *Thy* is a practically stable reference, its absolute values and physiological changes during sleep vary on the order of only a few tenths of a degree between and within animals. This criterion does not apply to active sleep as a result of the paradoxical vasomotion induced by the alteration of the autonomic regulation. Note that selective brain cooling (*Tp-Thy*) is negatively correlated (r= -0.607; b= -0.0067; P<0.05) with systemic heat exchanger vasoconstriction (*Thy-Ts*) during both quiet wakefulness and quiet sleep. Data from [17]

ischemia of systemic heat exchangers should bring the carotid and vertebral blood to the same temperature [11]. Moreover, the hydrostatic load affects selective and systemic brain cooling in opposite and quantitatively different ways [17]. A decrease in the negative hydrostatic load or a positive hydrostatic load (head-down posture at or below the heart level, respectively) enhances selective brain cooling, but induces a subsequent increase in the tonic vasoconstriction of systemic heat exchangers weakening systemic brain cooling. This may be more fully explained if it is assumed that the vessels of selective heat exchangers (the carotid rete-venous plexus) are readily dilated by a rise in transmural pressure and are less affected by the vasoconstrictor sympathetic outflow than those of systemic heat exchangers. Thus, posturally induced vasodilation in selective heat exchangers increases both the heat loss from the carotid blood and the carotid blood supply to the circle of Willis [17].

Mechanisms Underlying Hypothalamic Temperature Changes During the Ultradian Sleep Cycle

The hypothalamic temperature decrease found during QS with respect to QW is an effect of a state-dependent decrease in vasoconstrictor sympathetic outflow to the systemic heat exchangers of the head inducing vasodilation and increased heat loss, as shown for example, by the increase in ear pinna temperature [9, 11]. Recent results, showing that the experimentally induced changes in heat loss from the ear pinna consistently affect the spontaneous hypothalamic temperature decrease during QS [11], clearly demonstrate the influence of systemic heat exchanger vasomotion on brain cooling [1, 11].

A study of the hypothalamic temperature changes produced by transient drops in carotid blood flow as the result of short (100 s) bilateral common carotid artery occlusion shows a clear rise in hypothalamic temperature during QW and QS [11]. Systemic baroreceptive reflexes and autoregulation of cerebral blood flow buffer such drops in carotid blood flow by increasing the flow of vertebral blood [18]. Since the latter is warmer than carotid blood, the hypothalamic temperature is raised. This rise in hypothalamic temperature does not depend on metabolic heat production, since hypothalamic temperature soon reaches a plateau on long lasting (300 s) bilateral common carotid artery occlusion [11]. In contrast, bilateral common carotid artery occlusion during AS scarcely enhances the spontaneous increase in hypothalamic temperature. This would indicate that, during AS, carotid flow is spontaneously decreased, and that such a decrease is buffered by an increase in the amount of vertebral blood flowing into the circle of Willis. Figure 1b shows schematically the behavioural state-dependent changes in carotid and vertebral artery blood flows which would occur spontaneously across the ultradian wake-sleep cycle in the cat [11].

Further considerations are helpful to understand the complex mechanisms involved in the remarkable hypothalamic temperature change during AS [11]. A phenomenon of cerebral blood "steal" by the expansion of extracerebral beds, which elicits a decrease of carotid blood flow into the circle of Willis almost as important as that occurring during short bilateral common carotid artery occlusion, is conceivable on the basis of the impaired autonomic regulation during AS [19, 20, 23]. Such a sudden hemodynamic alteration is clearly revealed by the opposite ear pinna temperature changes at the onset of AS at ambient temperatures below and at/above ambient thermal neutrality, respectively [5, 6]. An uncontrolled expansion of vascular beds is indicated indirectly in two ways. Ear pinna vasoconstriction reverses to vasodilation in a cold environment, as a result of the depression in vasoconstrictor autonomic outflow. In addition, ear pinna vasodilation reverses to vasoconstriction in a neutral or warm environment, resulting from a decreased transmural pressure equally dependent on the alteration of autonomic cardiovascular regulation [3, 5, 6, 20-23]. This is further supported by the fact that the spontaneous decrease in ear pinna temperature during AS, observed at neutral ambient temperature, almost completely masks the effect of bilateral common carotid artery occlusion on ear pinna temperature and is more marked than that induced by bilateral common carotid artery occlusion during QW or QS when autonomic cardiovascular regulation (anastomotic supply to the ischemic carotid bed) is not impaired as it is during AS [19-23].

Summarizing, (Fig. 1b), a decrease in cool carotid blood supply and an increase in warm vertebral blood supply to the circle of Willis explain the hypothalamic temperature rise and its plateau during AS in species which have a carotid rete [11]. Only when systemic arterial blood temperature, including vertebral blood temperature, has been excessively lowered by QS-dependent heat loss in a very cold environment does this mechanism fail to raise the hypothalamic temperature during AS in cats [6].

When these considerations are taken together they may also help to explain the brain and hypothalamic temperature changes in species lacking a carotid rete (e.g., rabbit and rat).

In the rabbit, for instance, the basal portion of the brain, including the circle of Willis and the hypothalamus, is conductively cooled (selective cooling) by the cool venous blood draining from the nasal mucosa into the pterygoid plexus and the ophthalmic sinus [12]. Nevertheless, the same changes in hypothalamic temperature and systemic heat exchanger temperature during AS which are observed in cats [5, 6] are also found in rabbits [3] and rats [2]. The hemodynamic alterations in the common carotid bed during AS are also likely to impair this mechanism of selective brain cooling (deceleration of arterial carotid blood flow to the systemic heat exchangers and the circle of Willis, and of the venous blood flow from the former to the pterygoid and ophthalmic plexuses) and to increase the supply of vertebral blood to the circle of Willis. The increased perfusion of the brain by vertebral blood in the presence of depressed conductive cooling would be sufficient to raise hypothalamic temperature rapidly in rabbits since, also in this species, vertebral blood is warmer than the carotid blood perfusing the hypothalamus [1].

On the other hand, the morphofunctional prevalence of the internal carotid blood supply in primates [10], due to the high degree of telencephalization, provides unfavorable hemodynamic conditions, particularly in humans [24], for a significant enhancement of vertebral blood perfusion of the hypothalamus during AS. In the case of primates, carotid blood temperature is probably the most important determinant of hypothalamic temperature not only during QS but also during AS. In fact, systemic heat exchanger vasodilation during AS in the monkey has been shown to be consistently associated with a decrease in hypothalamic temperature [1, 25] which is in contrast to that found for cats [5, 6], rabbits [3] and rats [2].

Sympathetic and Postural Influences on Brain Cooling During the Ultradian Wake-Sleep Cycle

The behavioural states of QW, QS and AS are characterized by the interaction of sympathetic and postural influences on systemic and selective brain cooling [17]. During QW, the high tonic vasoconstrictor sympathetic outflow to systemic heat exchangers and the head-up posture (negative hydrostatic load decreasing transmural pressure) reduce both systemic and selective brain cooling: vertebral and carotid artery blood temperatures are higher than during QS and their difference is decreased [17]. From QW to QS, the state-dependent decrease in tonic vasoconstrictor sympathetic outflow [9] and the head-down posture (decrease in negative hydrostatic load and the raising of transmural pressure) increase systemic and selective brain cooling: vertebral and carotid blood temperatures are lower than during QW and their difference is increased [17]. In fact, a sharp drop in hypothalamic temperature, which is steeper at low ambient temperatures, is observed in cats as soon as the head is lowered to assume the sleep posture [4].

The influence of changes in posture on hypothalamic temperature is also fairly evident during AS [17]. However, in this state, the autonomically regulated balance between transmural pressure and vasoconstriction is disturbed in several vascular beds [3, 5, 6, 20-22]. The resulting alteration in the arterial blood perfusion of the circle of Willis, which receives an increased supply of vertebral blood to replace a decrease in the supply of carotid blood, weakens the influence of selective brain cooling on hypothalamic temperature which rises to approach the same temperature as the vertebral blood (Fig. 3) [11]. Therefore, during AS, hypothalamic temperature is primarily influenced by the temperature of vertebral blood which, as described previously, is dependent on the heat loss from systemic heat exchangers. Paradoxical changes (from the viewpoint of thermoregulation), as a result of the AS-dependent alteration in sympathetic regulation [20], occur in systemic heat exchanger vasomotion during AS, that is vasoconstriction and vasodilation at or above and below ambient thermoneutrality, respectively [2, 3, 5, 6]. At or above ambient thermoneutrality, the increase in hypothalamic temperature during AS is consistent with the systemic heat exchanger vasoconstriction occurring in response to a decrease in transmural pressure [5, 6, 11, 17]. In concomitance with the impairment of the mechanism of selective brain cooling during AS, this vascular response may additionally influence hypothalamic temperature by raising systemic arterial blood temperature (including vertebral blood temperature) at a rate which is dependent on the thermal inertia of the whole body. However, a hypothalamic temperature increase during AS is observed even at low ambient temperatures [2-6] in spite of the systemic heat exchanger vasodilation (increasing systemic heat loss) due to the state-dependent suppression of thermoregulatory vasoconstriction [20]. Also in this case, the effect of the impairment of selective brain cooling on hypothalamic temperature is associated with the effect of the thermal inertia of the whole body on the increased systemic heat loss. In fact, vertebral blood cooling during the short duration of AS is attenuated since the only possible influence on the temperature of vertebral blood comes from the

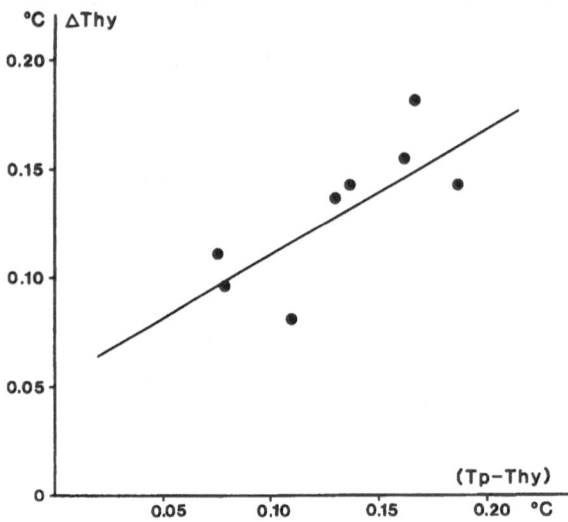

Fig. 3. *Changes in the rise in hypothalamic temperature (ΔThy) in active sleep as a function of the difference (Tp-Thy) between temperature of the pons (Tp) and Thy in quiet sleep just before the onset of active sleep.* Note that the *Thy* rise is positively correlated (r=0.783; b=0.718; P<0.05) with (*Tp-Thy*), that is, with the intensity of selective brain cooling in quiet sleep (data from [17]). The increased supply of vertebral blood to replace a decrease in the supply of carotid blood to the circle of Willis ("carotid-vertebral flow-shift"), weakens the influence of selective brain cooling on *Thy*. Thus, the larger (*Tp-Thy*), the more *Thy* rises to approach *Tp* during active sleep

long systemic venous return to the heart [11]. Consequently, the increase in vertebral blood supply to the circle of Willis during AS [11] fails to raise hypothalamic temperature only when systemic arterial blood temperature, including vertebral blood temperature, has been greatly lowered by excessive heat loss during the previous QS state in a very cold environment [6].

Conclusions

In general terms, hypothalamic temperature changes are the direct result of an imbalance between heat production and heat clearance. However, changes in the hypothalamic metabolic rate do not consistently underlie the temperature variations during the ultradian sleep cycle because of a tight blood flow-metabolism coupling, which also entails a heat clearance-heat production coupling [1]. Thus, the changes in hypothalamic temperature reflect only behavioural state-dependent adjustments in extracerebral circulatory variables influencing the temperature and flow of the arterial blood perfusing the brain. The hypothalamic temperature decrease during QS is mainly due to the increased heat loss at systemic and selective heat exchangers, whereas its increase during AS is the result of a state-dependent "carotid-vertebral flow-shift" in hypothalamic blood supply. Such events, however, scarcely affect the hypothalamic temperature since arterial blood temperature changes are efficiently buffered, at no energy cost, by the thermal inertia of the mass of body water [26]. Thus, a tight thermoregulatory control of hypothalamic temperature during sleep is unnecessary at ambient temperatures within or close to the thermoneutral range. This condition is advantageous in terms not only of energy saving but also of unhampered promotion of the centrally induced stereotyped adjustments of physiological functions during sleep. In fact, changes in hypothalamic temperature might activate thermoregulatory mechanisms which interfere with sleep processes, but only when ambient temperature deviates largely from the thermoneutral range [27].

References

1. Hayward JN, Baker MA (1969) A comparative study of the role of the cerebral arterial blood in the regulation of brain temperature in five mammals. Brain Res 16:417-440
2. Alföldi P, Rubicsek G, Cserni G, Obàl Jr F (1990) Brain and core temperatures and peripheral vasomotion during sleep and wakefulness at various ambient temperatures in the rat. Pflügers Arch 417:336-341
3. Franzini C, Cianci T, Lenzi P, Guidalotti PL (1982) Neural control of vasomotion in rabbit ear is impaired during desynchronized sleep. Am J Physiol 243:R142-R146
4. Parmeggiani PL, Agnati LF, Zamboni G, Cianci T (1975) Hypothalamic temperature during the sleep cycle at different ambient temperatures. Electroencephalogr Clin Neurophysiol 38:589-596
5. Parmeggiani PL, Zamboni G, Cianci T, Calasso M (1977) Absence of thermoregulatory vasomotor responses during fast wave sleep in cats. Electroencephalogr Clin Neuro-physiol 42:372-380
6. Parmeggiani PL, Zamboni G, Perez E, Lenzi P (1984) Hypothalamic temperature during desynchronized sleep. Exp Brain Res 54:315-320
7. Altman PL, Dittmer DS (1966) Environmental biology, F.A.S.E.B., Bethesda, MA
8. Franzini C (1992) Brain metabolism and blood flow during sleep. J Sleep Res 1:3-16
9. Azzaroni A, Parmeggiani PL (1995) Synchronized sleep duration is related to tonic vasoconstriction of thermoregulatory heat exchangers. J Sleep Res 4:41-47
10. Edvinsson L, MacKenzie ET, McCulloch J (1993) Cerebral Blood Flow and Metabolism, Raven, New York
11. Azzaroni A, Parmeggiani PL (1993) Mechanisms underlying hypothalamic temperature changes during sleep in mammals. Brain Res 632:136-142
12. Caputa M, Kadziela W, Narebski J (1976) Significance of cranial circulation for the brain homeothermia in rabbits. II. The role of the cranial venous lakes in the defence against hyperthermia. Acta Neurobiol Exp 36:625-638
13. Cabanac M (1986) Keeping a cool head. News Physiol Sci 1:41-44
14. Cabanac M, Caputa M (1979) Natural selective cooling of the human brain: evidence of its occurrence and magnitude. J Physiol (London) 286:255-264
15. Jessen C, Kuhnen G (1992) No evidence for brain stem cooling during face fanning in humans. J Appl Physiol 72:664-669
16. Nielsen B, Jessen C (1992) Evidence against brain stem cooling by face fanning in severely hyperthermic humans. Pflügers Arch 422:168-172

17. Azzaroni A, Parmeggiani PL (1995) Postural and sympathetic influences on brain cooling during the ultradian wake-sleep cycle. Brain Res 671:78-82
18. Busija DW, Heistad DD (1984) Factors involved in the physiological regulation of the cerebral circulation. Rev Physiol Biochem Pharmacol 101:162-211
19. Parmeggiani PL (1984) Autonomic nervous system in sleep. Exp Brain Res Suppl 8:39-49
20. Parmeggiani PL (1994) The autonomic nervous system in sleep. In: Kryger MH, Roth T, Dement WC (eds) Principles and practice of sleep medicine. Saunders, Philadelphia, pp 194-203
21. Cianci T, Zoccoli G, Lenzi P, Franzini C (1990) Regional splanchnic blood flow during sleep in the rabbit. Pflügers Arch 415:594-597
22. Cianci T, Zoccoli G, Lenzi P, Franzini C (1991) Loss of integrative control of peripheral circulation during desynchronized sleep. Am J Physiol 261:R373-R377
23. Mancia G, Zanchetti A (1980) Cardiovascular regulation during sleep. In: Orem J, Barnes CD (eds) Physiology in sleep. Research topics in physiology, vol 3. Academic Press, New York, pp 1-55
24. Hale AR (1960) Circle of Willis - Functional concepts, old and new. Am Heart J 60:491-494
25. Hayward JN, Baker MA (1968) Role of cerebral arterial blood in the regulation of brain temperature in the monkey. Am J Physiol 215:389-403
26. Parmeggiani PL (1995) Hypothalamic homeothermy across the ultradian sleep cycle. Arch Ital Biol 134:101-107
27. Parmeggiani PL (1987) Interaction between sleep and thermoregulation: An aspect of the control of behavioural states. Sleep 10:426-435

Part II

Pathophysiological Aspects of Human Sleep

Circadian Rhythm of Body Core Temperature in Neurodegenerative Diseases

G. Pierangeli, P. Cortelli, F. Provini, G. Plazzi and E. Lugaresi

Introduction

Diurnal variations in body temperature have been evident since the eighteenth century [1, 2]. A century later Gierse [3] and Ogle [4] documented 24-hour oscillations of body temperature in man. Simpson and Galbraith [5] were the first to measure the circadian rhythm of core body temperature (CRT) of animals placed under conditions of constant illumination or darkness, and to observe that the rhythm persisted unaltered suggesting the endogenous nature of the oscillations. The endogenous nature of circadian rhythms (CR) was confirmed by studying the golden hamster with a genetic mutation that produced a very short circadian rhythm of activity (a period of 20 instead of 24 h) [6].

Chronobiological studies often utilize body core temperature (BcT) for the study of CR since it is one of the most robust biological parameters, well protected by homeostatic processes [7]. The study of CRT in controlled conditions with concomitant monitoring of the sleep-wake cycle gives important information on central autonomic structures implicated in the control of this parameter. It is therefore a powerful tool for evaluating the involvement of these structures in neurodegenerative diseases with and without dysautonomia.

This paper summarizes the pattern of CRT in normal subjects, in multiple system atrophy (MSA) and in fatal familial insomnia (FFI). The study of CRT in these neurodegenerative diseases, the first characterized by autonomic hypofunction and the second by sympathetic hyperfunction, both idiopathic and of central origin, reveals the usefulness of this biological parameter as a marker of autonomic nervous system (ANS) function.

The Circadian Rhythm of Core Body Temperature

In normal human beings under natural conditions of lighting and social interaction, awakening at 07.00 h and retiring at 23.00 h, BcT cycles from the nightime low of approximately 36.5°C about 3 h before wakening (04.00 h), reaches 37.2°C by 09.00 h, continues to rise to a peak of 37.4°C at about 20.00 h, and falls thereafter to

Istituto di Neurologia, Università di Bologna, Via Ugo Foscolo 7, 40123 Bologna, Italy

reach the initial level of 36.5°C at 04.00 h [8]. In the human infant, the CRT is very weak or absent at birth and improves progressively during the first year of life; it is not clear yet at what age the amplitude reaches the adult level [9, 10]. In elderly subjects maintained without external time cues, a shortening of the period of CRT has been reported. Differences in the phase of the CRT between young and old humans, with possible gender interactions were also observed. Finally, a reduction in the amplitude of the CRT in old age was observed in humans [10]. When human subjects are maintained under conditions of constant illumination and without access to external time cues, their CRT "free runs", usually with a period of about 25 h. If the subjects are allowed to sleep or take naps whenever they wish, most sleep takes place during the low part of CRT, but a significant bout of sleep (early afternoon nap) is also found around the time of the CRT peak [11, 12]. More recent evidence suggests an endogenous period closer to 24 h for humans [7]. The circadian system has an inherent flexibility that allows free-running rhythms to become synchronized to the numerous environmental, behavioral, and social cues that provide the structure of daily life [7]. The natural alternation between light and dark provides the most important synchronizing cue for the mammalian circadian system. However, artificial light-dark schedules can entrain rhythms only if the schedule does not deviate substantially from the intrinsic circadian rhythm of the organism. Other factors that have been shown to entrain circadian rhythms include food availability, ambient temperature, forced activity and social cues. The exogenous synchronizers or "zeitgebers" may affect the expression of circadian rhythms without affecting the underlying pacemaker generating the rhythm. Such transient effects are called "masking effects" [7]. Light may have a direct rising effect on BcT more evident during the rest period of the rest-activity cycle [13]. Ambient temperature has a minimal effect on BcT in homeothermic animals but can determine major changes in the strategies for maintaining BcT by changing levels of heat production and/or heat loss [14]. Although BcT typically is elevated at the phase when feeding occurs, the thermodynamic action of food does not play a major role in the normal diurnal temperature variation [14]. Subjects fasting for 48 h continue to show CRT [11], but a reduction in metabolic rate and hence heat production accompanies prolonged severe energy restriction; a suppression of thermoregulation via thermogenesis may accompany dieting [15]. Although vigorous muscular activity can elevate body temperature, the normal rest-activity cycle has relatively little influence on the CRT [14].

The CRT lost some of its adaptive significance after the evolution of endothermy. However if we consider the CRT together with other biological rhythms like the sleep-wake cycle and other autonomic phenomena, it is evident that all these 24-hour oscillations tend to lower the level of energy expenditure during the rest period that is useful in terms of physiological homeostasis.

CRT and Thermoregulation

The major central mechanism of the thermoregulatory system is located in the hypothalamus in the preoptic and anterior areas where thermal information from central and peripheral thermoreceptors is integrated and from which efferent thermoregulatory discharges arise. The balance between heat production and heat loss determines the body temperature. Thermoregulatory effector mechanisms in humans consist of shivering and non-shivering thermogenesis for heat production, and cutaneous vasomotor response and sweating for heat dissipation. Heat exchange between the body and the environment occurs through radiation, conduction and convection. The rate of heat loss is regulated by adjusting cutaneous blood flow and sweating [16].

Aschoff suggested that the circadian oscillation of BcT is a regulated process associated with an oscillation of the thermoregulatory set point [17]. The concept of set point derives its appeal from the convenient way in which it explains the physiological processes involved in the phenomenon of fever. But whether the CRT itself results from a change in set point similar to that observed during fever remains controversial. In principle, there is no reason why the CRT should result from a change in set point. An oscillatory process could very well be produced by a decrease in heat production or an increase in heat loss independent of the thermoregulatory set point. In order to preserve a rhythm of body temperature despite the fluctuations in heat production determined by the activity rhythm, the thermoregulatory system more probably relies on mechanisms of heat loss [9]. It seems that an early nocturnal subcutaneous hyperaemic phase is the basis for the later BcT fall. This delay might be ascribed to the relatively large heat capacity of the central part of the human organism. The nocturnal subcutaneous hyperaemia is probably a centrally elicited active vasodilatation [18].

The Circadian System Controlling the CRT

The circadian timing system can be divided into three major components: clock, input pathways and output pathways. In mammals, the master oscillator or clock (circadian pacemaker) is located in the suprachiasmatic nucleus (SCN). Virtually all SCN neurons may be circadian oscillators and there must be some mechanism for coupling or synchronizing the activity of these oscillators [19]. Principle input pathways belong to the circadian visual system for the transmission of photic information and are represented by the retinohypothalamic tract (RHT) and by the geniculohypothalamic tract (GHT) [19]. A third serotoninergic input pathway originates from the midbrain and seems to be important in the transmission of the zeitgeber effect of locomotion [19]. There appear to be two distinct output mechanisms from the SCN both modulated by neuronal activity: (a) a primarily intrahypothalamic projection system, utilizing classical synaptic transmission, and (b) a potentially global system allowing secretion or diffusible output transmitters into the cerebrospinal fluid (CSF) or cerebral vasculature [19]. There are four major SCN-efferent projections in the hamster. These include (a) an anterior projection to the anterior paraventricular thalamus, ventral lateral septum, and bed nucleus of stria terminalis; (b) a generalized periventricular fiber system innervating much of the medial hypothalamus from the preoptic region to the premammillary area; (c) a lateral thalamic projection to the intergeniculate leaflet (IGL); and (d) a posterior projection to the posterior paraventricular thalamus, precommissural nucleus, and olivary pretectal nucleus [19].

The recent demonstration of a retinal clock in hamsters [20] has renewed an old question: "Do mammals have a single clock or multiple oscillators?" There are many controversial studies demonstrating the persistence of the CRT in SCN-lesioned animals [21-24] that seem to indicate the existence of an unknown oscillator regulating the 24-hour variations of this rhythm. Against this hypothesis is the case of a man with a discrete metastasis in the ventral hypothalamus and optic chiasm who developed an abnormal CRT without alteration of the 24-hour mean BcT [25]. At the present time there is not enough experimental evidence to confirm the hypothesis that the CRT and rest-activity cycle are controlled by two distinct circadian pacemakers.

Interactions Between CRT and Sleep-Wake Cycle

The CRT is not a simple consequence of the daily activity cycle, since paralyzed or bedridden subjects, as well as normal subjects kept inactive in bed, have an ongoing cycle of BcT independent of levels of muscular activity [26]. The magnitude of the sleep-related falls in BcT which are considered "masking" components of the CRT depends on the phase of the CRT and is maximal in the descending phase. Daily fluctuations in temperature continue during sleep deprivation [27, 28].

The CRT in humans has strong influences on both timing and duration of sleep. When, in environments devoid of time cues, the sleep-wake and BcT cycles uncouple and "free run" with different circadian periodicities, the voluntary bedtime and sleep onset are more likely to occur near the BcT nadir [29]. In these subjects sleep parameters correlated with CRT and not with length of prior wakefulness. Several human studies suggest that sleep propensity and REM sleep propensity may have a bimodal distribution in relation to the CRT. This interaction of sleep and BcT rhythms has important implications for several clinical and applied problems [30].

BcT influences sleep but, on the other hand, thermoregulatory phenomena are strictly state-dependent. In fact downward regulation of BcT is a specific feature of NREM sleep. This shows a state-dependent change in the sympathetic outflow controlling heat loss and production. However thermoregulatory autonomic responses to thermal loads are still normally operative during NREM sleep. On the other hand, thermoregulation is practically suspended during REM sleep and BcT is more influenced by ambient temperature. This could result from a transient and reversible inactivation of the hypothalamic structures during REM sleep leading to an impairment of the sympathetic outflow [31,32]. In all species thermoregulatory and other autonomic phenomena of NREM sleep suggest the presence of closed loop operations that maintain homeostasis at a lower level of energy expenditure compared with wakefulness. In contrast the autonomic activity during REM sleep is characterized in all species by widescale variability because of the occurrence of nonhomeostatic (open loop) operations [33].

The Role of ANS in Thermoregulation

In mammals, thermoregulatory responses to changes in body or ambient temperatures are controlled by hypothalamic-preoptic integrative mechanisms that drive subordinate brainstem and spinal somatic and autonomic mechanisms. The ANS is a fundamental agent in the control of BcT. In particular its sympathetic division regulates heat loss and conservation at the level of heat exchangers by inducing thermoregulatory vasomotion, piloerection and (in humans) sweating. This autonomic division also regulates metabolic heat production by skeletal muscle (shivering thermogenesis) and catabolic catecholaminergic mechanisms (non-shivering thermogenesis of brown adipose tissue) [33].

The central control of thermoregulation, as other autonomic, neuroendocrine and behavioral responses critical for homeostasis, is regulated by the central autonomic network (CAN) that is an internal regulation system of the brain [34]. The CAN is involved in visceromotor, neuroendocrine, complex motor and pain modulating control mechanisms essential for adaptation and survival. The CAN consists of a group of interconnected areas of the telencephalon, diencephalon and brain stem which control preganglionic sympathetic and parasympathetic visceromotor outputs. The components of the CAN include (a) the insular and medial prefrontal cortices, (b) the central nucleus of the amygdala and the bed nucleus of the stria terminalis, (c) the anterior ventral and mediodorsal thalamic nuclei and the hypothalamus, (d) the periacqueductal gray matter in the midbrain, (e) the parabrachial Kölliker-Fuse region in the pons, (f) the nucleus of the tractus solitarius (NTS) and (g) the medullary intermediate reticular zone, particularly the ventrolateral medulla [34].

The CAN is critical for (a) tonic background excitation to autonomic and respiratory motoneurons, (b) coordination of spinal preganglionic units, (c) reflex adjustments of cardiovascular and other autonomic responses and (d) integrated autonomic, neuroendocrine and behavioral responses for maintenance of homeostasis, emotional expression and responses to stress. These functions depend on important features of the CAN, including reciprocal interconnections, parallel organisation, state-dependent activity and neurochemical complexity [34]. Reciprocal interconnection allows continuous feedback interactions and integration of autonomic responses, and is mediated by the median forebrain bundle and the dorsal longitudinal fasciculus. Parallel organization is required because the central autonomic control depends on the activity of several parallel pathways, rather than on specific autonomic "centers" as classically described. The central autonomic control is state-dependent in the sense that tonic, reflex and integrative control of autonomic function depends on the physiologic and behavioral state of the individual. Thus it can be affected by respiration, sleep-wake cycle, emotional state, attention and other factors [34].

The CRT as a Marker of ANS Function

The CAN is likely to regulate circadian variations in BcT translating and transmitting to the effectors the 24-hour oscillations generated by the biological clock. Therefore the study of CRT in controlled conditions may reflect the function of the circadian oscillator and of these central autonomic structures that effectively produce the daily BcT oscillations.

Methodological Aspects

For an accurate evaluation of CRT, continuous recording of BcT by means of a portable device (Mini-logger) gives detailed information on the rhythm. The sleep-wake cycle can be monitored at the same time by an ambulant polygraphic recorder (Oxford Medilog 9200) recording electroencephalogram (EEG: C3-A2, C4-A1), electro-oculogram (EOG), electrocardiogram (ECG) and electromyogram (EMG). For good control of exogenous factors the patients live in a temperature- and humidity-controlled room (24 ± 1°C, 40-50% humidity) during the study, lie in bed except for eating, and are exposed to a controlled light-dark schedule (dark period: 23.00 h - 07.00 h). The diet can be controlled by placing the subjects on a 1800 kcal/day diet divided into three meals (08.00 h, 12.00 h, 18.00 h) and three snacks (10.00 h, 16.00 h, 22.00 h). A habituation period of 48 h in the same room and with the light-dark schedule and the diet described is recommended before the start of the recording session.

For the analysis of rhythmicity, the evaluation of the time series for BcT with the single cosinor method [35] by means of a computerized procedure [36] is a useful technique. This procedure determines whether or not there is a rhythm within a 24 h period (p< 0.05) and evaluates the following parameters with their 95% confidence limits: (a) mesor (midline estimating statistic of rhythm): rhythm-adjusted 24 h average; (b) amplitude: difference between the maximum value measured at the acrophase and the mesor in the cosine curve; (c) acrophase: lag between reference time (24.00 h) and time of highest value of the cosine function used to approximate the rhythm.

The sleep stages can be scored according to the criteria suggested by Rechtschaffen and Kales [37] considering epochs of 2 min to synchronize sleep-wake evaluation and temperature data. Sleep parameters such as total sleep time (TST), percentage of sleep time from 22.00 h to 07.00 h, and duration and percentage with respect to TST of each sleep phase can be evaluated applying an automatic sleep stager corrected by hand scoring.

The CRT in Neurodegenerative Diseases

Multiple System Atrophy

Classification, Pathology and Clinical Aspects

The term multiple system atrophy (MSA) first used by Graham and Oppenheimer in 1969 [38] refers to a sporadically occurring adult-onset neurodegenerative disease previously described variously as striatonigral degeneration (SND), olivopontocerebellar atrophy (OPCA) or Shy-Drager syndrome (SDS) according to the clinical predominance of parkinsonism, cerebellar signs or autonomic failure, respectively [39-41]. Autonomic failure (AF) may also occur in a "pure" form without any other neurological signs (PAF), previously called "idiopathic orthostatic hypotension" [42].

Neuropathologically, MSA is characterized by cell degeneration and gliosis in the striatum substantia nigra, Purkinje cell layer of the cerebellar cortex, pontine nuclei and inferior olives. In almost all patients with AF, whether they had PAF or MSA, there was found up to 80% reduction in the number of sympathetic preganglionic neurons in the intermediolateral cell columns of the spinal cord. Brainstem abnormalities have also been noted including the dorsal nucleus of the vagus, the locus coeruleus and the nucleus tractus solitarii.

The clinical presentation [43] in patients with MSA with AF is highly variable, at least at the onset of the disorder: patients may present with parkinsonism poorly responding to levodopa, with a cerebellar syndrome, with progressive AF (postural hypotension, defective sweating, failure of erection in males, constipation), or with a variable combination of these syndromes. The prognosis of MSA with AF is usually determined by the progress of neurologic features over the course of about six years.

Circadian Rhythms

The sleep-wake organization is frequently abnormal in MSA patients. The more frequently reported alterations are sleep fragmentation, and reduced TST and REM sleep with shortened REM sleep latency [44, 45]. MSA with AF patients presented a consistent circadian trend in blood pressure that is the inverse of the normal pattern, with the highest pressure at night and the lowest in the morning [45,46].On the contrary, MSA patients without AF had a normal blood pressure circadian variation [45]. This strongly suggests that normal blood pressure day-night changes depend on the proper functioning of the sympathetic nervous system. The urinary excretion rate of 6-hydroxymelatonin, the major metabolite of melatonin, was impaired in patients with MSA who excreted abnormally low amounts of the metabolite, or relatively high amounts during the day with the lowest excretion during the interval between midnight and 06.00 h [47]. The authors suggested an impairment of the central pathways controlling periodical oscillations in the synthesis and in the excretion of melatonin. The CR of plasma melatonin was found to be markedly abnormal in MSA patients irrespective of the presence of sympathetic failure, suggesting the involvement of central rather than peripheral autonomic mechanisms [48, 49]. A patient with MSA with AF showed nocturnal polyuria associated with an abnormal circadian rhythm of plasma antidiuretic hormone (ADH) secretion with increased levels during the day and without the physiological increase during the night, suggesting neurodegenerative involvement of the hypothalamus-posterior hypophysis system [50].

Cardiovascular Autonomic Function

The cardiovascular autonomic function alterations found in five patients with probable MSA with AF submitted to CRT study were typical of this neurodegenerative disease. The patients (4 men and 1 woman aged 63±8 years) showed severe symptomatic postural hypotension and had normal noradrenaline (NA) plasma levels at supine rest (274±181 pg/ml; n.v.= 194±106 pg/ml) without a significant increase at orthostatic stress (312±201 pg/ml; n.v.= 374±140 pg/ml). The other physiologic tests (Valsalva manoeuvre, deep breathing, isometric handgrip) in patients with MSA with AF showed a sympathetic efferent defect of central origin. The parasympathetic efferent lesion was less severe than was the sympathetic one. One MSA with AF patient showed a significant impairment of sudomotor function evaluated by means of Guttmann's method [51].

The Circadian Rhythm of Body Core Temperature

The mean BcT 24-hour oscillations of four patients with probable MSA (1 male and 3 females aged 62±4 years), of five patients with probable MSA with AF (4 males and 1 female aged 63±8 years) and of nine sex- and age-matched controls are presented in Fig. 1.

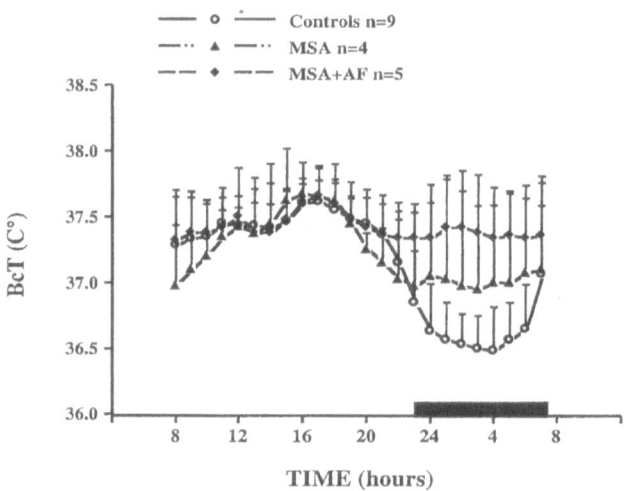

Fig. 1. *Circadian oscillations of body core temperature (BcT) in multiple system atrophy (MSA) with autonomic failure (MSA + AF) patients and controls. Each point is the mean ± SD of one hour's BcT values (1 sample each 2 min). The black bar on x axis shows the dark period. MSA with AF patients present an absence and MSA patients a reduction in BcT physiological nocturnal fall*

Analysis of rhythmicity showed the presence of a significant BcT rhythm in all patients. The mesor was higher and the amplitude reduced with respect to the control group in two out of four MSA patients and in two out of five MSA with AF patients. Three out of five MSA with AF patients showed a shift in the acrophase.

Comparing the two groups of patients (with and without AF) with the controls, the mesor and the amplitude of the patients with AF were significantly different, the mesor being increased and the amplitude reduced. The MSA group (without AF) showed the same trend, but the changes in rhythm parameters did not reach a significant level (Fig. 2).

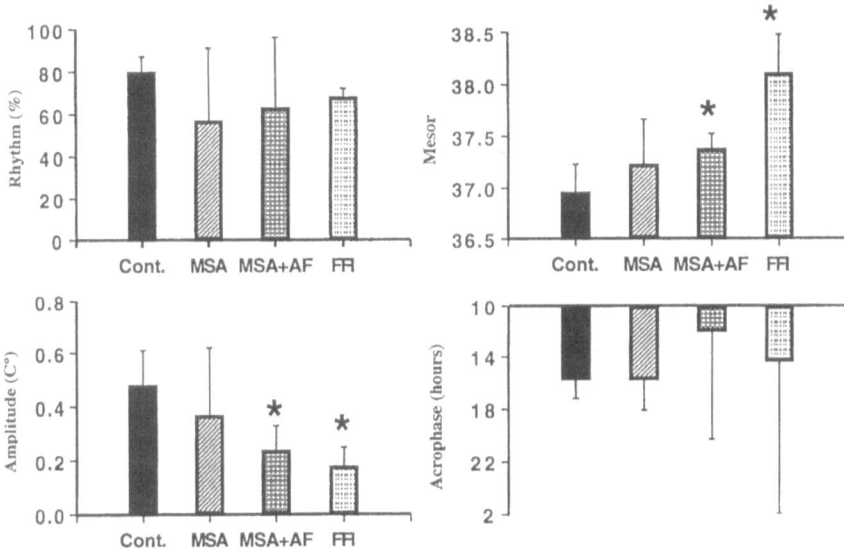

Fig. 2. *Circadian rhythm parameters (rhythm percentage, mesor, amplitude and acrophase) in multiple system atrophy (MSA), MSA with autonomic failure (MSA + AF), fatal familial insomnia (FFI) patients and controls.* Statistically significant differences (<0.05) are indicated by *

Sleep-Dependent Changes in Body Core Temperature

Nighttime sleep was more disrupted in MSA patients than in controls, with a decrease in total sleep time and in the percentage of slow wave sleep (phase 3-4) but with a normal percentage of REM sleep. In the control group, BcT showed the expected nocturnal reduction during NREM sleep with the minimum during REM sleep when thermoregulation is suspended. In patients with AF this pattern was absent, and rectal temperature was significantly higher during NREM and REM sleep. In MSA patients without AF, BcT was significantly higher only during REM sleep (Fig. 3).

The impairment of the CRT in 50% of patients with MSA, regardless of the presence of AF, indicates that the abnormalities observed are not only the effect of an impaired control of the cardiovascular system. However, the presence of AF is associated with more severe abnormalities in BcT temperature rhythm. The

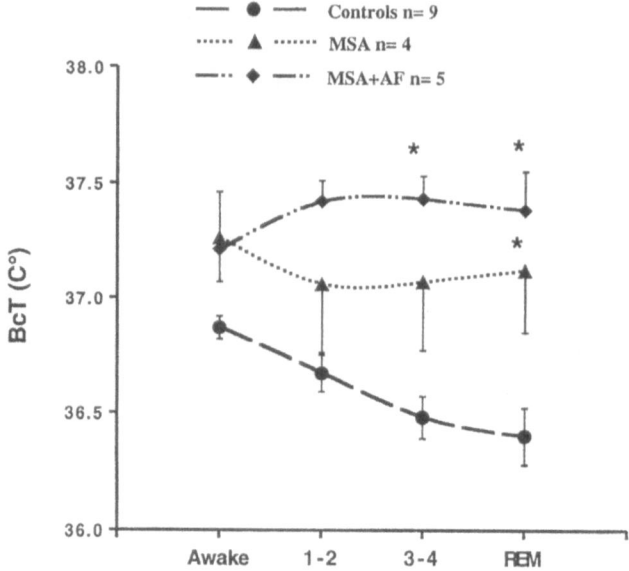

Fig. 3. *Mean BcT (body core temperature) during wakefulness and during different sleep stages in multiple system atrophy (MSA), MSA with autonomic failure (MSA + AF) patients and controls.* Statistically significant differences (<0.05) are indicated by *. MSA with AF patients present a mean BcT higher than controls in stage 3-4 and in REM sleep. In MSA patients the same alteration is significant only in REM sleep

CRT is abnormal mainly because during sleep the BcT does not fall. This indicates that the state-dependent mechanisms of autonomic regulation, that are the result of the integrated activity of the higher functions of the ANS, are impaired in MSA. In this case too the abnormalities were more pronounced in the group with AF. We cannot exclude that a primary dysfunction of the endogenous oscillator such as the SCN could be the cause of the altered CRT observed in MSA. Up to now no pathological abnormalities of this nucleus have been described. Recently Nakamura et al. [52] described a reduction of large-sized histaminergic neurons in the tuberomammillary nucleus of the hypothalamus in patients with MSA. This finding raises interesting questions about the influence that this lesion could have in the regulation of the sleep-wake cycle and of CR in MSA.

Fatal Familial Insomnia

Clinical, Pathologic and Genetic Aspects

Fatal familial insomnia (FFI) is an autosomal dominant disease clinically characterized by untreatable insomnia, dysautonomia and motor signs [53, 54]. The histopathological hallmark of FFI is a severe atrophy of the thalamus, especially of the anterior ventral and mediodorsal nuclei, associated with a variable involvement of the inferior olive, striatum and cerebellum [55]. In addition, a mild to moderate spongiform degeneration is seen in the cerebral cortex in subjects with the longest duration of symptoms. Biochemical [56], genetic [57] and transmission [58, 59] studies reveal that FFI is a transmissible prion disease linked to a mutation of codon 178 of the prion protein gene (PRNP), leading to the substitution of aspartic acid with asparagine (178^{Asn}). The coding region of the PRNP has a polymorphism at codon 129 that results in two variant alleles, one coding for methionine (129^{Met}) and the other for valine (129^{Val}) [60]. FFI is invariably linked to the presence of the methionine codon at position 129 of the mutant allele [61]. On the basis of the Met/Val polymorphism at codon 129 of the nonmutated allele, however, FFI subjects can be further divided into homozygotes ($129^{Met/Met}$) or heterozygotes ($129^{Met/Val}$). The course of the disease is longer in the heterozygotes at codon 129.

Circadian Rhythms

In some patients 24-hour polysomnographic recordings demonstrated an almost total absence of sleep patterns detectable on the EEG from the early stage of the disease. Only short episodes of abnormal REM sleep occurred without the physiological loss of muscle tone lasting a few seconds or minutes. Patients with a more prolonged course of the disease had progressively shortened sleep cycles, with a progressive reduction in total sleep time but the cyclic sleep organization and the gradual transition from one stage to another was lost from the onset. Spindle activity and K-complexes were totally absent or degraded from the onset of the disease in all patients. This was reflected in the disordered sudden passage from one phase of sleep to another, an expression of the loss of regulation of the intrinsic cyclic NREM and REM sleep activity. Polygraphic analysis showed consistent elevation of the values of blood pressure, heart rate and BcT in FFI without the natural circadian oscillations [45, 53].

Somatotropin (GH), prolactin (PRL), cortisol, corticotropin (ACTH), plasma catecholamines (noradrenaline and adrenaline), and melatonin were assayed at 30 min intervals. A dysfunction of the pituitary-adrenal axis was prominent in all cases with elevated serum cortisol and normal levels of ACTH [62]. Catecholamines displayed increased levels throughout the 24 hours considerably synchronized with heart rate. However, complete rhythm obliteration for these hormones occurred in the terminal stage of FFI. Plasma melatonin concentrations gradually decreased as the disease progressed. A significant circadian rhythm was detected in the earlier recordings, but disappeared in the most advanced stage [63]. The physiological nocturnal elevation of GH was absent simultaneously with sleep loss, whereas a significant 24-hour component of variations in serum PRL levels was present for months after total disruption of the sleep-wake cycle with normally placed nocturnal acrophase [64]. Complete rhythm abolition was also observed later for PRL.

In summary, the advanced stages of the disease are invariably characterized by the disappearance of any circadian rhythmicity of the hormones evaluated.

Cardiovascular Autonomic Function

An imbalanced autonomic control with preserved parasympathetic activity and a higher background and stimulated sympathetic activity was the conclusion derived from the autonomic tests [54]. This sympathetic overactivity was evident in the elevated noradrenaline plasma levels at rest which increased further under orthostatic stress, the exaggerated blood pressure responses to physiological stimuli (postural change, Valsalva manoeuvre, isometric handgrip), the absent blood pressure response to noradrenaline infusion, the increased heart rate response to atropine, the diminished depressor and sedative effects of clonidine. We confirmed that sympathetic hyperactivity is a hallmark of FFI by evaluating a larger number of patients at an earlier stage of the disease by means of power spectral analysis of heart rate variability [65].

The Circadian Rhythm of Body Core Temperature

The CRT was evaluated in three patients with FFI. Case 1(129[Met/Val]) was a 36-year-old woman at the onset of the disease which lasted 35 months and who underwent CRT recording at 15, 17, 20, 22 and 23 months. Case 2 (129[Met/Val]) was a 58-year-old woman at the onset of the disease which lasted 32 months, and who was studied at 6, 9 and 14 months. Case 3 (129[Met/Met]) was a 53-year-old man with a duration of the disease of 8 months evaluated at 4 and 6 months.

The 24-hour BcT and sleep-wake cycle oscillations showed progessive alterations with the advancing of the disease (Figs. 4, 5). A significant 24 h oscillation

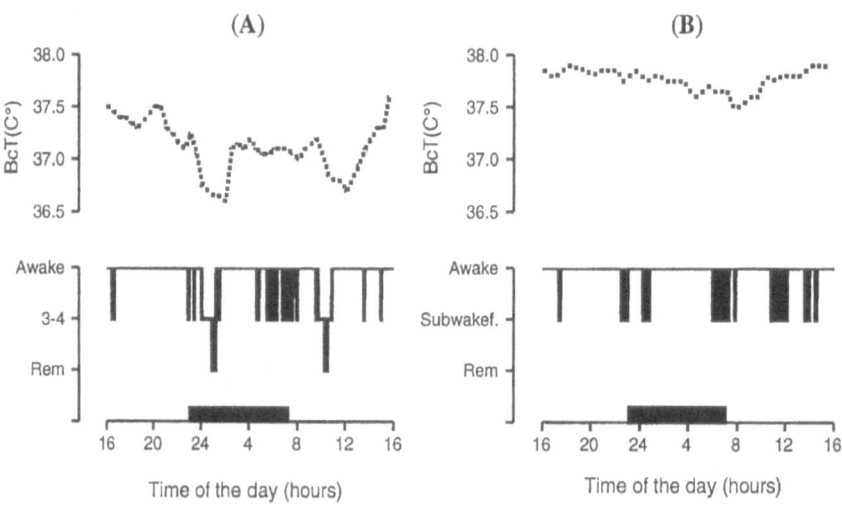

Fig. 4. *Progressive alterations of circadian oscillations in BcT (body core temperature) in a patient with fatal familial insomnia (FFI, case 1, 129[Met/Val]) at the 15th (A) and at the 23rd (B) month of the disease.* The simultaneous EEG recording disclosed the persistence of normal state-dependent behavior of BcT. In the *BcT curve* each point is the mean ± SD of one hour's values (1 sample each 30 min). The *black bar* on x axis shows the dark period

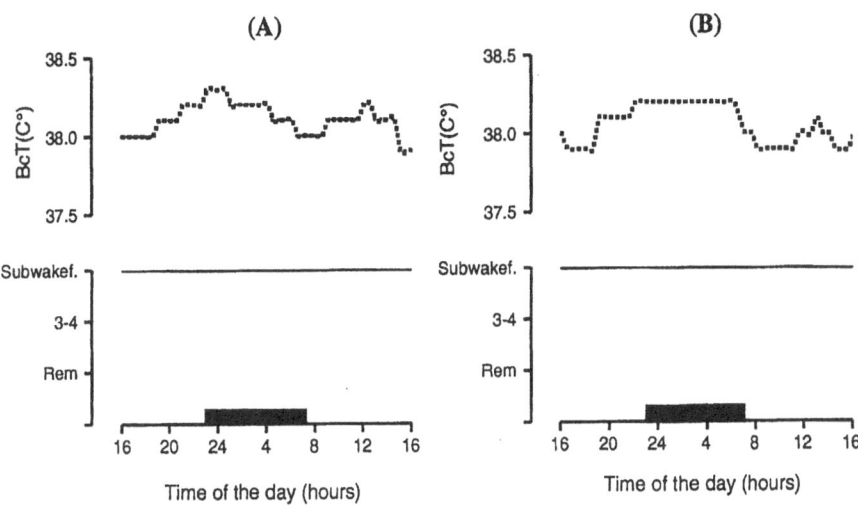

Fig. 5. *Alteration of circadian oscillations in BcT (body core temperature) in a patient with fatal familial insomnia (FFI, case 3, 129^{Met/Met}) at the 6th (A) and 9th (B) months of the disease.* The simultaneous EEG recording shows a condition of total agrypnia in both studies. In the *BcT curve* each point is the mean ± SD of one hour's values (1 sample each 30 min). The *black bar* on x axis shows the dark period

in BcT was present in all recordings. The mesor was higher with respect to the control range in all study sessions except in the first two studies of patient 1. The amplitude was significantly reduced in all observations except one, and the acrophase was shifted in the last two studies of patient 2 and in all evaluations of patient 3. The comparison between FFI and control groups, selecting the last study of each FFI patient, showed a significant increase in the mesor and reduction of the amplitude. The mesor was higher in the FFI group with respect to the MSA group (Fig. 2). In FFI the evaluation of mean BcT in different sleep stages was impossible due to marked sleep alterations. However, the first polygraphic recordings in patients 1 and 2 showed short episodes of NREM sleep accompanied by a sudden reduction in BcT (Fig. 4).

A 24-hour oscillation persisted in FFI patients but profound alterations of the CRT always appeared with the disease progression. The mesor progressively increased concurrent with decrease in the acrophase; the acrophase was shifted in most studies. These data must be considered together with the alterations observed in all autonomic and neuroendocrine functions. Clinicopathologic correlations demonstrate the role of the thalamus as an integrative neural structure placed between the limbic system and the hypothalamus controlling the homeostatic balance of the organism. A lesion in the visceral part of the thalamus could release the hypothalamus from inhibitory cortical control, resulting in a permanently activated nervous system: a sort of brain unable to rest [66]. Loss of sleep, persistent basal overactivation, and hyperactivity of the sympathetic nervous system all fit within the hypothesis that ergotrophic functions prevail over trophotrophic functions [67]. Thus the progressive increase in mesor and the respective decrease of amplitude seem to be determined by the prevalence of neurogenic mechanisms for heat production on those responsible for heat dissipation that try to balance the system with a relative hyperactivity. Therefore in those cases where short episodes of NREM persist, BcT, blood pressure and heart rate fall abruptly but still display a normal state-dependent behavior. From the chronobiologic point of view, FFI offers a unique opportunity to observe the daily patterns of autonomic and hormonal parameters in human subjects independent of sleep-wake cycle. The persistence of an albeit impaired CRT after months of total agrypnia confirms the presence of an endogenous oscillator regulating BcT circadian variations independently of the sleep-wake cycle.

Concluding Remarks

The study of CRT in controlled experimental conditions is a sensitive and reproducible method to evaluate ANS function, mainly in its central component. The normal functioning of the CAN is crucial for physiological circadian oscillations of BcT when there is no impairment of the endogenous oscillator.

The same CRT alterations are present in neurodegenerative diseases associated with autonomic hypo- or hyperfunction, suggesting that this method is not specific for a particular CAN dysfunction but is a global functionality index.

References

1.	De Gorter J (1736) De perspiratione insensibili. Batavorum, Vander
2.	Hunter J (1778) On the heat of animals and vegetables. Philos Trans R Soc Lond (Biol) 68:7-49
3.	Gierse A (1842) Quoniam sit ratio caloris organici. Halle, Doctoral Thesis
4.	Ogle JW (1866) On the diurnal variations in the temperature of the human body in health. St George's Hosp Rep 1:220-245
5.	Simpson S, Galbraith JJ (1906) Observations on the normal temperature of the monkey and its diurnal variation, and on the effect of changes in the daily routine on this variations. Trans Royal Soc Edinburgh 45:65-104
6.	Ralph MR, Menaker M (1988) A mutation of the circadian system in golden hamsters. Science 241:1225-1227
7.	Murphy PJ, Campbell SS (1996) Physiology of the circadian system in animals and humans. Clin Neurophysiol 13:2-16
8.	Aschoff J, Gerecke U, Wever R (1967) Phasenbeziehungen zwischen den circadiaen perioden der aktivitat und der kerntemperatur beim menschen. Pflugers Arch 295:173-183
9.	Refinetti R, Menaker M (1992) Circadian rhythm of body temperature. Physiol Behav 51:613-637
10.	Glotzbach SF, Edgar DM, Boeddiker M, Ariagno RL(1994) Biological rhythmicity in normal infants during the first 3 months of life. Pediatrics 94:482-488

11. Wever RA (1979) The circadian system of man. Results of experiments under temporal isolation. Springer-Verlag, New York
12. Campbell SS, Zulley J (1989) Evidence for circadian influence on human slow wave sleep during daytime sleep episodes. Psychophysiology 26:580-585
13. Fuller CA, Sulzman FM, Moore-Ede MC (1979) Circadian control of thermoregulation in the squirrel monkey (*Saimiri sciureus*). Am J Physiol 236:R153-R161
14. Moore-Ede MC, Sulzman FM, Fuller CA (1982) The clocks that time us. Physiology of the circadian timing system. England, Harvard University Press
15. Karklin A, Driver S., Buffenstein R (1994) Restricted energy intake affects nocturnal body temperature and sleep patterns 1-3. Am J Clin Nutr 59:346-9
16. Ogawa T, Low PA (1993) Autonomic regulation of temperature and sweating. In: Low PA (ed) Clinical autonomic disorders. Little Brown, London, pp 79-91
17. Aschoff J (1970) Circadian rhythm of activity and body temperature. In: Hardy JD, Gagge AP, Stolwijk JAJ (eds) Physiological and behavioral temperature regulation. Charles C Thomas, Springfield, IL, pp 905-919
18. Sindrup JH, Petersen LJ, Madsen SM, Kristensen JK, Kastrup J (1995) Nocturnal temperature and subcutaneous blood flow in humans. Clin Physiol 15:611-622
19. Miller JD, Morin LP, Shwartz WJ, Moore RY (1996) New insights into the mammalian circadian clock. Sleep 19:641-667
20. Tosini G, Menaker M (1996) Circadian rhythms in cultured mammalian retina. Science 272:419-420
21. Dunn JD, Castro AJ, McNulty JA (1977) Effect of suprachiasmatic ablation on the daily temperature rhythm. Neurosci Lett 6:345-348
22. Powell EW, Pasley RN, Brockway B, Scheving LE, Lubanovic W, Halberg F (1977) Suprachiasmatic dinuclear lesion alters circadian temperature rhythm's amplitude and timing in light-dark synchronized rats. Chronobiologia 4:270
23. Nakayama T, Hardy JD (1969) Unit responses in the rabbit's brain stem to changes in brain and cutaneous temperature. J Appl Physiol 27:848-857
24. Fuller CA, Lydic R, Sulzman FM, Albers HE, Tepper B, Moore-Ede MC (1981) Circadian rhythm of body temperature persists after suprachiasmatic lesions in the squirrel monkey. Am J Physiol 241:R385-R391
25. Schwartz WJ, Busis NA, Hedley-Whyte ET (1986) A discrete lesion of vertebral hypothalamus and optic chiasm that disturbed the daily temperature rhythm. J Neurol 233:1-4
26. Kleitman N (1963) Sleep and wakefulness, 2nd edn. University of Chicago Press, Chicago
27. Patrick GTW, Gilbert JA (1896) On the effects of loss of sleep. Psychol Rev 3:469-483
28. Kreider MB (1961) Effects of sleep deprivation on body temperature. Fed Proc 20:214
29. Glotzbach SF, Heller HC (1994) Temperature regulation. In: Kryger MH, Roth T, Dement WC (eds) Principles and practice of sleep medicine, 2nd edn. WB Saunders, Philadelphia pp 260-275
30. Richardson GS, Malin HV (1996) Circadian rhythm sleep disorders: pathophysiology and treatment. J Clin Neurophysiol 13:17-31
31. Parmeggiani PL (1980) Temperature regulation during sleep: A study in homeostasis. In: Orem J, Barnes CD (eds) Physiology in sleep. Research topics in physiology. Academic Press, New York, pp 97-143
32. Parmeggiani PL (1987) Interaction between sleep and thermoregulation: An aspect of the control of behavioral states. Sleep 10:426-435
33. Parmeggiani PL (1994) The autonomic nervous system in sleep. In: Kryger MH, Roth T, Dement WC (eds) Principles and practice of sleep medicine, 2nd edn. WB Saunders, Philadelphia, pp 194-203

34. Benarroch EE (1993) The central autonomic network: Functional organization, dysfunction, and perspective. Mayo Clin Proc 68:988-1001
35. Halberg F, Reinberg A (1967) Rythmes circadiens et rythmes de basses fréquences en physiologie humaine. J Physiol (Paris) 59 (suppl):117-200
36. Mojon A, Fernandez JR, Hermida RC (1994) ChronoLab: an interactive software package for chronobiologic time series analysis written for the Macintosh computer. Bioengineering and Chronobiology Laboratories, E.T.S.I. Telecomunicacion, University of Vigo, Spain
37. Rechtschaffen A, Kales A (1968) A manual of standardized terminology, techniques and scoring system for sleep stages of human subjects. Publication 204. US Public Health Service
38. Graham JG, Oppenheimer DR (1969) Orthostatic hypotension and nicotine sensitivity in a case of multiple system atrophy. J Neurol Neurosurg Psychiatry 32:28-34
39. Adams RD, Van Bogaert L, Van der Eecken H (1961) Dégénerescences nigro-striées et cérébello-nigro-striées. Psychiatr Neurol 142:219-259
40. Shy GM, Drager DA (1960) A neurological syndrome associated with orthostatic hypotension. Arch Neurol 2:511-527
41. Dejerine J, Thomas AA (1900) L'atrophie olivo-ponto-cérébelleuse. Nouv Iconog Salpetrière 13:330-370
42. Bradbury S, Eggleston C (1925) Postural hypotension. A report of 3 cases. Am Heart J 1:73-86
43. Wenning GK, Ben Shlomo Y, Magalhaes M, Daniel SE, Quinn P (1994) Clinical features and natural history of multiple system atrophy. An analysis of 100 cases. Brain 117:835-845
44. Plazzi G, Corsini R, Provini F, Pierangeli G, Martinelli P, Montagna P, Lugaresi E, Cortelli P (1997) Rem sleep behavior disorders in multiple system atrophy. Neurology 48:1094-1097
45. Cortelli P, Pierangeli G, Provini F, Plazzi G, Lugaresi E (1996) Blood pressure rhythms in sleep disorders and dysautonomia. In: Portaluppi F, Smolensky MH (eds) Time-dependent structure and control of arterial blood pressure. The New York Academy of Sciences, New York, pp 204-201
46. Mann S, Altman DG, Raftery EB, Bannister R (1983) Circadian variation of blood pressure in autonomic failure. Circulation 68:477-483
47. Tetsuo M, Polinsky RJ, Markey SP, Kopin IJ (1981) 6-Hydroxymelatonin excretion in patients with orthostatic hypotension. Endocrinol Metab 53:607-610
48. Vaughan GM, McDonald SD, Jordan RM, Allen JP, Bell R, Stevens EA (1979) Melatonin, pituitary function and stress in humans. Psychoneuroendocrinology 4:351-362
49. Cortelli P, Portaluppi F, Pierangeli G, Maltoni P, Pavani A, Lugaresi E (1996) Impaired circadian rhythm of melatonin in multiple system atrophy. Neurology 46 (Suppl.2):A311
50. Ozawa T, Tanaka H, Miyatake T, Tsuji S (1993) Shy-Drager Syndrome with abnormal circadian rhythm of plasma antidiuretic hormone secretion and urinary excretion. Internal Medicine 32:225-227
51. Guttmann L (1947) The management of the quinizarin sweat test (Q.S.T.). Postgrad Med J 23:353-366
52. Nakamura S, Ohnishi K, Nishimura M, Suenaga T, Akiguchi I, Kimura J, Kimura T (1996) Large neurons in the tuberomammillary nucleus in patients with Parkinson's disease and multiple system atrophy. Neurology 46:1693-1696
53. Lugaresi E, Medori R, Montagna P, Baruzzi A, Cortelli P, Lugaresi A, Tinuper P, Zucconi M, Gambetti P (1986) Fatal familial insomnia and dysautonomia with selective degeneration of thalamic nuclei. New Engl J Med 315:997-1003

54. Cortelli P, Parchi P, Contin M, Pierangeli G, Avoni P, Tinuper P, Montagna P, Baruzzi A, Gambetti P, Lugaresi E (1991) Cardiovascular dysautonomia in fatal familial insomnia. Clin Auton Res 1:15-21

55. Manetto V, Medori R, Cortelli P, Montagna P, Tinuper P, Baruzzi A, Rancurel G, Hawn JJ, Vanderhaegen JJ, Mailleux P, Bugiani O, Tagliavini P, Bouras C, Rizzuto N, Lugaresi E, Gambetti P (1992) Fatal familial insomnia: clinical and pathologic study of five new cases. Neurology 42:312-319

56. Monari L, Chen SG, Brown P, Parchi P, Petersen RB , Mikol J, Gray F, Cortelli P, Montagna P, Ghetti B, Goldfarb LG, Gajdusek DC, Lugaresi E, Gambetti P, Autilio-Gambetti L (1994) Fatal familial insomnia and familial Creutzfeldt-Jacob disease: different prion proteins determined by a DNA polymorphism. Proc Natl Acad Sci USA 91:28-39

57. Medori R, Tritschler HJ, Leblanc A, Villare F, Manetto V, Chen HY, Xue R, Montagna P, Cortelli P, Tinuper P, Avoni P, Mochi M, Baruzzi A, Hawn JJ, Lugaresi E, Autilio-Gambetti L, Gambetti P (1992) Fatal familial insomnia: a prion disease with a mutation at codon 178 of the prion protein gene. New Engl J Med 326:444-487

58. Tateishi J, Brown P, Kitamoto T, Hoque ZM, Roos R, Wollman R, Cervenakova L, Gajdusek DC (1995) First experimental transmission of fatal familial insomnia. Nature 376:434-435

59. Collinge J, Palmer MS, Sidle KCL,Gowland I, Medori R, Ironside J, Lantos P (1995) Transmission of fatal familial insomnia to laboratory animals. Lancet 346:569-570

60. Collinge J, Palmer MS, Dryden AJ (1991) Genetic predisposition to iatrogenic Creutzfeldt-Jakob disease. Lancet 337:1441-1442

61. Goldfarb LG, Peterson RB, Tabaton M, Brown P, Leblanc A, Montagna P, Cortelli P, Julien J, Vital C, Pendelbury WW, Haltia M, Wills PR, Hauw JJ, McKeever PE, Monari L, Schrank B, Swergold GD, Autilio-Gambetti L, Gajdusek PE, Lugaresi E, Gambetti P (1992) Fatal familial insomnia and familial Creutzfeldt-Jakob disease: disease phenotype determined by a DNA polymorphism. Science 258:806-809

62. Portaluppi F, Cortelli P, Avoni P, Vergnani L, Contin M, Maltoni P, Pavani A, Sforza E, Degli Uberti EC, Gambetti P, Lugaresi E (1994) Diurnal blood pressure variation and hormonal correlates in fatal familial insomnia. Hypertension 23:569-576

63. Portaluppi F, Cortelli P, Avoni P, Vergnani L, Maltoni P, Pavani A, Sforza E, Degli Uberti EC, Gambetti P, Lugaresi E (1994) Progressive disruption of the circadian rhythm of melatonin in fatal familial insomnia. J Clin Endocrinol Metab 78:1075-1078

64. Portaluppi F, Cortelli P, Avoni P, Vergnani L, Maltoni P, Pavani A, Sforza E, Manfredini R, Montagna P, Roiter I, Gambetti P, Fersini C , Lugaresi E (1995) Dissociated 24-hour patterns of somatotropin and prolactin in fatal familial insomnia. Neuroendocrinology 61:731-737

65. Cortelli P, Pierangeli G, Parchi P, Barletta G, Contin M, Gambetti P, Montagna P, Lugaresi E (1994) Power spectral analysis reveals sympathetic hyperactivity as an early feature of fatal familial insomnia (FFI). Neurology 44 (suppl 2):A363

66. Lugaresi E (1992) The thalamus and insomnia. Neurology 42 (suppl.6):28-33

67. Hess WR (1969) Hypothalamus and thalamus: Experimental documentation. Georg Thieme, Stuttgart

Descriptive Epidemiology of Excessive Daytime Sleepiness

R. D'Alessandro, R. Rinaldi, L. Vignatelli and C. Tonon

Methodologic Considerations

Daytime sleepiness is the common experience of the tendency to sleep during the daytime. When this phenomenon is not invalidating, is justified by lifestyle and is sporadic, it may be considered a normal expression of our needs. If, however, the tendency to fall asleep during the day becomes excessive, undesired, inappropriate, disturbing and persistent, it becomes pathologic and may be referred to as *excessive daytime sleepiness* (EDS) [1-3]. EDS is being studied with increasing interest, since it is indicative of several diseases and has major familial and social effects.

The subjective experience of EDS can be identified by the "Multiple Sleep Latency Test" (MSLT), which provides quantitative and qualitative information on the tendency to sleep during the daytime [2, 4-6]. As this test cannot be used on large samples of the general population for reasons of cost and time, the descriptive epidemiology of EDS must of necessity rely on questionnaires where the presence or absence of the symptom is investigated. The use of different questions in the questionnaire, even if they resemble each other, can give rise to differences in the answers. This highlights the importance of the linguistic aspect and of the difficulty of translating definitions into other languages [7-9]. Moreover, a question about a symptom, and not about a specific disease, inevitably introduces variables linked to personal experience, education and the impact of the disturbance on the familial and social environment.

The questionnaire should therefore be prepared bearing these points in mind so as to satisfy the criteria of intelligibility, reproducibility and validity [10]. The validation will disclose the sensitivity and specificity of the questions employed against the test considered diagnostic of the phenomenon under investigation (in this case the MSLT) and is necessary to establish the reliability of the screening tool and hence the results obtained [11]. In addition to a clear explanation of how to prepare the questionnaire, the modality of its administration and, in the case of interviews, the type of training for interviewers should both be shown. Finally, the type of population on which the survey is to be carried out must be specified,

Unità di Epidemiologia, Istituto di Neurologia, Università di Bologna, Via Ugo Foscolo 7, 40123 Bologna, Italy as well as the way the sample is to be obtained, the participation rate and the characteristics of non-participants. Only by satisfying all these points, will it be possible to compare and generalize the results.

Bearing these guidelines on descriptive epidemiology in mind, we reviewed the literature on the prevalence of EDS in the general population and investigated the phenomenon in the city of Bologna.

Review of the Literature

We considered the descriptive epidemiologic studies of EDS on non-selected samples of the general population [12-19], also including two surveys which excluded elderly subjects [20, 21]. Moreover, studies carried out on a single sex [22-26] or age group [8, 22-36] were examined, as long as the criteria of non-selection was assured. In some studies the prevalence rate was not calculated, as this was not part of the principal aims of the study, but it was obtainable from other results reported [12, 25, 29, 30, 32]. Studies carried out on selected populations were excluded, such as volunteers, certain groups of workers, subjects undergoing regular medical check-ups and in-patients.

The studies examined reported widely differing prevalence rates of EDS in the general population (varying from 0.3 to 36%). This wide variability of rates is probably due mainly to the use of inhomogeneous methodologic criteria, such as the definition of EDS, the way of gathering the data and the type of population studied [37]. These points will be analyzed one by one.

Definition of EDS

The difference between the definitions used in questionnaires is the first point to be considered. In some studies the subjective experience of EDS is explored, employing questions with phrases such as "feel sleepiness/feel sleepy" [8, 14, 22, 25, 27, 34], "feel tired" or "with a compulsive desire to sleep" [28], "struggling to stay awake" [27, 33], "being so sleepy that you have to take a nap" [31, 36]. Other questionnaires study the objective consequences of EDS, i.e. the occurrence of falling asleep during the day in a non-physiologic way, and use definitions such as "the excessive tendency to fall asleep" [20], "being troubled by falling asleep" [15], falling asleep "when you did not intend to" [17], or falling asleep "against your will", "where you do not want to", "even if you don't want to, you can't help it" [25, 30, 35], or in different situations [18, 25, 33]. Broman et al. [21] define EDS by the concomitant presence of the subjective experience of "sleepiness" and of "involuntary falling asleep" during the day. Because of the difficulties to find the best word to indicate EDS and to avoid overestimating the phenomenon, Billiard et al. [23] looked for "daytime sleep episodes", considering them as the culminating expression of sleepiness. "Sleeping too much" and "too much sleep", without a defined reference to events/experiences during the day, were investigated by Karacan et al. [12], Bixler at al. [13], and Ford and Kamerow [16]. However, we underline that questions on "sleep episodes" and "sleeping too much/too much sleep" did not strictly investigate the EDS.

The wide spectrum of the described definitions is enhanced by reference to current or previous symptoms and to specifications on the frequency and circumstances of the disturbances. In addition, some studies use more than one definition, usually with reference to different situations of occurrence of the disturbance, knowing that the object of the study is highly subjective [18, 25, 33]. Finally, the Austrian study by Zeitlhofer et al. [19] and several north European surveys do not provide a precise definition of the disturbance [24, 26, 29, 32], even if the "Basic Nordic Sleep Questionnaire" [7] was devised on the basis of the latter studies.

Administration of the Questionnaire

Another element of disparity within the surveys is the way the questionnaire is administered. The administration can take place through interviews, either by phone [17, 20, 33] or directly [14, 16, 18, 25, 30, 31, 34-36], or by self-administration [8, 15, 21, 24, 26, 28, 29, 32]. Billiard et al. [23] used a supervised completion of the form while Ancoli-Israel et al. [27] carried out interviews partly on the phone and partly at home. In some studies the modality of administration is not clearly specified [12, 13, 19, 22].

Validation

The numerous types of disparity described could be overcome to some extent by validating the questionnaire against a test considered diagnostic for EDS, such as the MSLT. This procedure was used in the preparation of questionnaires such as the "Epworth Sleepiness Scale" (ESS) [38], the "Pittsburg Sleep Quality Index" (PSQI) [39] and the "Sleep-Wake Activity Inventory" (SWAI) [40], which have not yet been used in epidemiologic studies on the general population. Among the studies considered here, only Schmidt-Nowara and collegues' investigation validated the questionnaire [18].

Population Studied

The surveys were carried out on a general population resident in defined geographic areas. Only in some studies was the population taken from such sources as population registers [21, 32], voter registration lists [35], telephone directories [17, 27, 33], households [12, 13, 15, 16, 18, 36], lists provided by general practitioners or by other types of health services [25, 30, 34], twin registers [8], or draftee lists [22, 23].

Few studies were carried out on the whole general population [12-21]. In the study by Lugaresi et al. [14] children were included, while adolescents were questioned in the investigations by Partinen and Rimpelä [20] and Zeitlhofer et al. [19]. Two studies, however, excluded the most elderly subjects [20, 21].

The rest of the surveys were carried out on sub-samples consisting only of males [22-25] or females [26], or representing the younger age range [22, 23], young-adults [32], adults [8, 24-26, 28, 30, 33], or the elderly [27, 29, 31, 34-36]. It should be pointed out that there is no exact correspondence regarding age range in any of the studies.

Choice and Size of Samples

Samples were taken from the general population at random [12, 13, 15-21, 24, 26-30, 32-36] and in a stratified manner [13, 15, 21, 28-30, 35]. In some studies the stratification, however, was not representative of the general population [21, 28-30]. Other studies considered representative [8, 14] or total samples [23, 25]. The sample of elderly subjects used by Foley et al. [31] was obtained from different geographic areas partly at random and partly from the total population [41].

The number of subjects taking part in the studies varied from 396 [21] to 58 162 [23]. The participation rate ranged from 23% in a study where the participants had to undergo a neurophysiologic examination [27], to 99.4% in a study carried out on draftees [22]. In some studies [13, 14, 16, 17, 19, 20, 33, 34], the absence of information regarding the initial sample size did not permit the calculation of the response rate or a comparison between the characteristics of the initial sample and those of the participants. Finally, only in the surveys by Ancoli-Israel et al. [27] and Jennum and Sjøl [30] were studies made on a sample of non-participants in order to estimate the reliability of the prevalence obtained among the participants. None of the studies, however, report an estimation of EDS frequency in a sample of non-participants.

EDS Prevalence Rates in the Literature

The wide variability of investigation tools and the different characteristics of the population samples led us to consider results with caution. The prevalence rates of EDS in the general population are shown in Table 1. When the rates are presented for age and sex, the disturbance tends to be higher in some cases in younger subjects [13, 14, 16] and to increase with age in others [15, 17, 20]. Data regarding sex also differ, being higher either among men [13, 15, 20] or women [16, 17, 21].

Figures 1-3 show the prevalence rates for sex and age ranges in all cases where the data was available; the age ranges were sufficiently comparable. The study by Bliwise and King [33] was excluded from the figures as no unambiguous definition of EDS was given. For the study by Stradling et al. [25] which used several definitions we chose the value regarding the question referring to "falling asleep during the day against your will", which is close to the definition of EDS used in other works [30, 35]. Generally speaking, the prevalence rate is higher among the elderly, with a tendency to higher values in men [15, 31, 34, 35]. Most EDS prevalence rates obtained in the adult population are around 10% with slightly higher values in women. In the young/young-adult group prevalence rates are more varied, with values covering a range halfway between those of the elderly and those of the adults. Within this group, studies by Klink and Quan [15] and Broman et al. [21]

Table 1. Epidemiological studies on the prevalence of EDS in the general population: both sexes and according to gender

Author	Year	Age	Prevalence rate: Both sexes	Prevalence rate: Men	Prevalence rate: Women
Karacan et al. [12]	1976	≥18	0.3%	n.a.[a]	n.a.
Bixler et al. [13]	1979	18-80	7.1% (current or previous) 4.2% (current)	4.3% (current)	4.1% (current)
Partinen & Rimpela [20]	1982	15-64	3%	3.4%	2.5%
Lugaresi et al. [14]	1983	3-94	8.7%	n.a.	n.a.
Klink & Quan [15]	1987	≥18	12%	12.3%	11.7%
Ford & Kamerow [16]	1989	≥18	3.2%	2.8%	3.5%
Phillips et al. [17]	1989	n.a.	36%	33%	39%
Schimdt-Nowara et al. [18]	1991	≥18	2.4%-13.3%	n.a.	n.a.
Zeitlhofer et al. [19]	1994	≥14	29%	n.a.	n.a.
Broman et al. [21]	1996	20-64	9%	7%	10%

[a] n.a. = not available

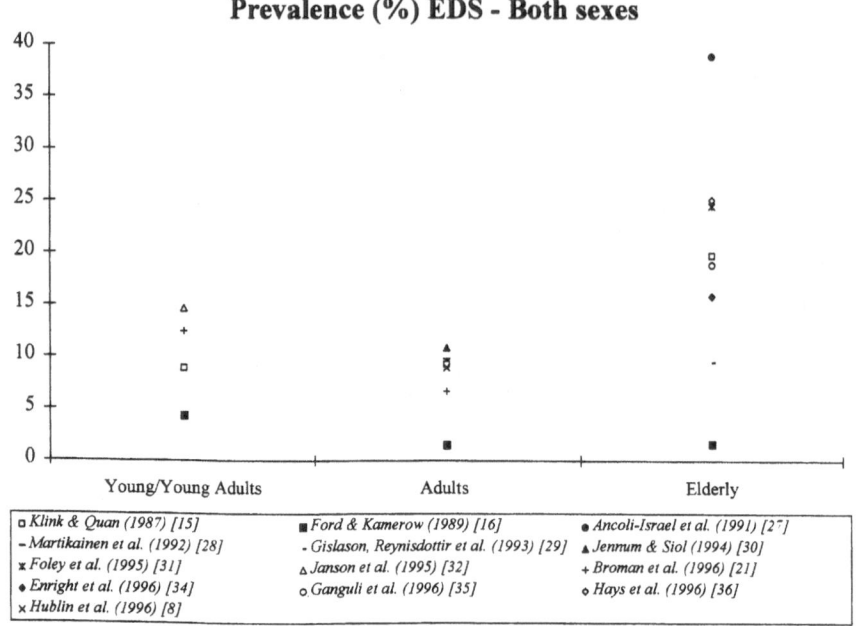

Fig. 1. *Prevalence rates of EDS in both sexes according to age: review of epidemiological studies*

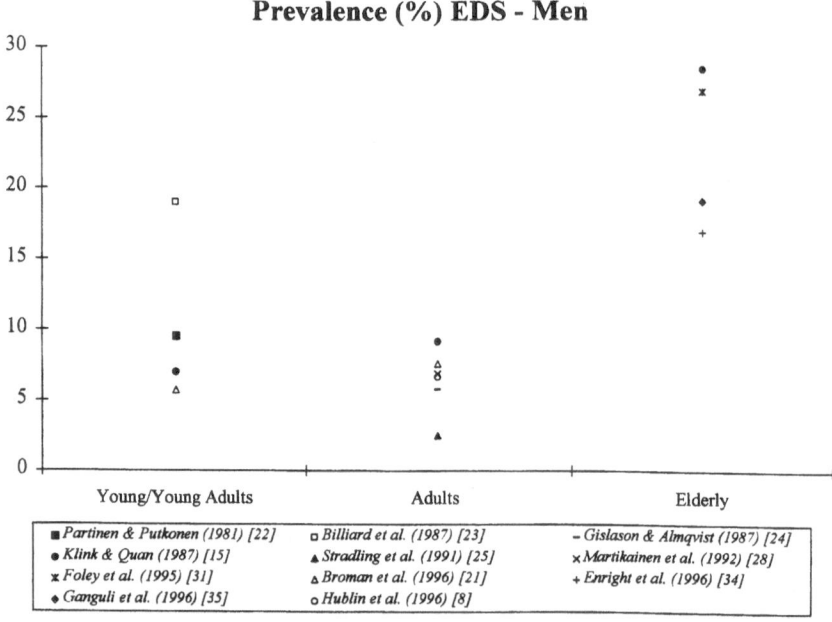

Fig. 2. *Prevalence rates of EDS in men according to age: review of epidemiological studies*

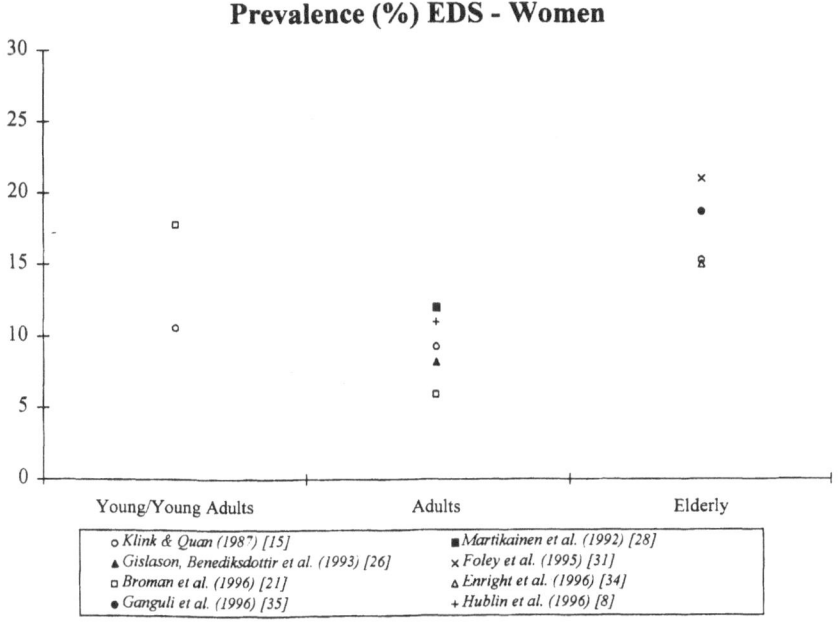

Prevalence (%) EDS - Women

Fig. 3. *Prevalence rates of EDS in women according to age: review of epidemiological studies*

investigated both sexes, and reported a higher prevalence in women than in men (10.6 vs. 7% and 17.8% vs. 5.7%, respectively).

EDS in the City of Bologna

Questionnaire

A self-administered questionnaire prepared by the Epidemiology Unit of the Institute of Clinical Neurology, Bologna University, was used for the survey. The questionnaire was devised following the guidelines given by Stone [10], taking into account the international literature on EDS and the clinical experience obtained at the Sleep Center in our institute. The validation of the questions against the result of the MSLT was carried out on a sample of subjects with and without the disorder before the start of the survey [42]; a high sensitivity and specificity (100% and 74%, respectively, for MSLT $\leq$ 5 min) was obtained for the question about irresistible sleepiness occurring at least almost every day in the last three months. The result of the validation was better when details of frequency were provided rather than the circumstances of occurrence. The questions regarding tiredness, resistible sleepiness and sudden sleep attacks, also included in the questionnaire, did not prove valid. On the basis of the validation results, we therefore defined EDS as the presence *in the last three months of the need or the desire to sleep during the day, being unable to resist and so falling asleep at least almost every day.* The response to this question subsequently was used for the calculation of the prevalence rates.

Table 2 lists the questions on different aspects of the phenomenon and the sub-items regarding frequency and circumstances of occurrence.

Table 2. Main questions and sub-items common to all of them. The sub-items A, B, C and D regard the frequency of the problem. E, F, G concern different situations of symptom occurrence

Main Questions	Sub-Items
1 In the last three months have you ever felt tired, weary or listless, or had difficulty in concentrating and in paying attention during the day?	**Frequency of occurrence** A Did it occur once or twice a month? B Did it occur once or twice a week? C Did it occur almost daily? D Did it occur every day?
2 In the last three months have you ever felt the need or the desire to sleep during the day, but were able to resist and not fall asleep?	**Situations of occurrence** E Did it occur after meals?
3 In the last three months have you ever felt the need or the desire to sleep during the day, and were unable to resist and so fell asleep?	F Did it occur during undemanding or repetitive situations, such as watching television, reading the newspaper, travelling by train, etc?
4 In the last three months have you ever fallen asleep suddenly and without warning during the day?	G Did it occur during demanding situations, such as eating, chatting with friends, working, riding a bicycle, etc?

Population and Procedure

The study was carried out in the city of Bologna from January 1993 to January 1995. The total population at the start of the study was 400 858 citizens (186 982 males and 213 876 females), 89% of whom were aged ≥ 18.

The questionnaire was mailed to a sample of 1509 subjects aged ≥ 18, obtained from the population register, at random and stratified for sex and age. Eighty-one subjects had died or were no longer at the same address and it was possible to substitute only 53 of these with the help of a supplementary list. Therefore the final sample consisted of 1481 subjects. Of these, 746 (50.4%) returned the form completed. The rest did not participate either because of lack of consent (no interest or time, mistrust in scientific research, fear of lack of respect for anonymity, suspicion of fraud, absence of payment) or because of inability to fill in the answers (illiteracy, dementia, serious psychiatric illness). In

Table 3. Whole population of the city of Bologna, initial sample, participants and a sample of non-participants interviewed by telephone, according to age and sex

Age (years)	Whole population n (%)		Initial sample n (%)		Participants n (%)		Telephone interview n (%)	
	Men	Women	Men	Women	Men	Women	Men	Women
18-29	34 048 (20.6)	31 089 (16.1)	135 (19.9)	123 (15.3)	82 (23.9)	86 (21.3)	11 (15.3)	8 (9.2)
30-39	27 797 (16.8)	26 728 (13.8)	118 (17.4)	109 (13.6)	72 (21.0)	76 (18.9)	10 (13.9)	7 (8.1)
40-49	25 933 (15.7)	28 151 (14.6)	103 (15.1)	110 (13.7)	53 (15.5)	60 (14.9)	10 (13.9)	13 (14.9)
50-59	27 205 (16.5)	31 722 (16.4)	109 (16.0)	128 (16.0)	60 (17.5)	71 (17.6)	11 (15.3)	12 (13.8)
60-69	25 739 (15.6)	33 143 (17.2)	104 (15.3)	139 (17.4)	41 (11.9)	63 (15.6)	13 (18.0)	16 (18.4)
≥70	24 308 (14.7)	42 290 (21.9)	111 (16.3)	192 (24.0)	35 (10.2)	47 (11.7)	17 (23.6)	31 (35.6)
Total	165 030 (100)	193 123 (100)	680 (100)	801 (100)	343 (100)	403 (100)	72 (100)	87 (100)

order to prevent the absence of the investigated disturbance from becoming an involuntary selection criteria as a result of non-participation, a sub-sample of 159 subjects (72 men and 87 women), equal to around 20% of non-participants for each sex and age, was interviewed by phone using the main questions. The characteristics of the population studied are shown in Table 3.

Data Analysis

The prevalence rate of EDS in the general population was calculated for the 746 subjects participating in the study. The prevalence was also calculated using the whole initial sample of 1481 subjects as denominator to estimate the minimum prevalence of EDS in our population. The analysis was then performed for both sexes and for different age groups. Finally, for each value the 95% confidence interval according to the Poisson distribution was calculated [43]. The EDS frequency was also obtained by phone in a group of non-participating subjects.

Results

The prevalence rates of EDS in the general population of the city of Bologna are reported in Table 4. The prevalence rate of EDS was 7.1% in the total population, 6.1% in men and 7.9% in women. The minimum prevalence was estimated to be 3.6% in the total sample, 3.1% in men and 4% in women. In the young group (between 18 and 29 years), women showed an EDS prevalence value twice as high as that in men, that is 8.2% compared to 3.7%. In the adult group (between 30 and 59 years), women showed fairly uniform values, ranging from 5.2% to 6.7%. Among the men, however, the values were inhomogeneous, ranging from rarity or absence of the disturbance in subjects in their 30s and 40s to a value of 8.4% in the group in their 50s. In the elderly (aged 60 years or over) prevalence values were higher than those in the rest of the population; within this group the men showed a peak at around 14% which was slightly higher than that of the women.

Table 4. Prevalence rates (%) of EDS (C.I. 95%) in the city of Bologna, according to age and sex

Age (years)	Both sexes	Men	Women
18-29	6.0 (2.9-11.0)	3.7 (0.8-10.8)	8.2 (3.3-16.9)
30-39	4.1 (1.5-8.9)	2.8 (0.3-10.1)	5.2 (1.4-13.3)
40-49	3.6 (0.9-9.2)	0 (0-3.0)	6.7 (1.8-17.2)
50-59	6.8 (3.1-12.9)	8.4 (2.7-19.6)	5.6 (1.5-14.3)
60-69	13.5 (7.4-22.7)	14.6 (5.4-31.8)	12.7 (5.5-25.0)
≥ 70	12.2 (5.9-22.5)	14.3 (4.6-33.3)	10.6 (3.4-24.7)
Total	7.1 (5.3-9.3)	6.1 (3.8-9.3)	7.9 (5.4-11.1)

Among all subjects interviewed by phone, the EDS frequency was 16.3% (9.7% in men and 21.8% in women). These results, even if considered with caution, show that the EDS rate of the non-participating group is not lower than that of the participants.

Conclusions

The EDS prevalence in the general population remains an open question, as shown by the wide methodologic disparity found in the existing literature. The most frequently reported rates range from 3% to 13.3% showing that EDS is a common problem. The analysis for sex and age reveals a higher prevalence in young women and an increased prevalence in the elderly, especially among men, suggesting that the phenomenon is linked to different causes, age and sex dependent.

A fundamental requisite for future epidemiologic studies and for the identification of risk factors is the formulation of an unambiguous internationally-accepted definition of EDS. We propose a validated, sensitive and specific questionnaire for the investigation of EDS in the general population [42]. The prevalence study which we performed in the city of Bologna using this tool was the first to use a valid and unambiguous definition of EDS. We therefore believe that the prevalence rate of 7.1% is quite near to realistic values. Even though we obtained a response rate of only 50%, a sample of non-participating subjects interviewed by phone did not show a lower prevalence rate. The rates obtained in the city of Bologna show that EDS tends to increase with age and that a frequency peak is present among young females, in agreement with some previous studies [15, 21].

The analysis of the clinical factors associated with EDS, such as the quality of nocturnal sleep, the presence of somatic symptoms, the use of drugs and the presence of snoring and nocturnal apneas is currently in progress.

Acknowledgements

This work was supported by C.N.R., "Progetto Finalizzato Invecchiamento" (No. 91.00379.40.115.02164), and partly by "Fondo per la Ricerca Sanitaria Finalizzata Regione Emilia-Romagna". Anne Collins revised the English text.

References

1. Diagnostic Classification Steering Committee, Thorpy MJ, Chairman (1990) International classification of sleep disorders: diagnostic and coding manual. American Sleep Disorders Association, Rochester, Minnesota
2. Roth T, Roehrs TA, Carskadon MA, Dement WC (1994) Daytime sleepiness and alertness. In: Kryger MH, Roth T, Dement WC (eds) Principles and practice of sleep medicine. WB Saunders Company, Philadelphia, pp 40-49

3. Culebras A (ed) (1996) Clinical handbook of sleep disorders. Butterworth-Heinemann, Boston, Massachussetts, USA
4. Carskadon MA, Dement WC, Mitler MM, Rota T, Westbrook PR, Keenan S (1986) Guidelines for the multiple sleep latency test (MSLT): a standard measure of sleepiness. Sleep 9:519-524
5. Carskadon MA (1993) Evaluation of excessive daytime sleepiness. Neurophysiol Clin 23:91-100
6. Thorpy MJ (1992) The clinical use of the Multiple Sleep Latency Test. Sleep 15:268-276
7. Partinen M, Gislason T (1995) Basic Nordic Sleep Questionnaire (BNSQ): a quantitated measure of subjective sleep complaints. J Sleep Res 4[Suppl 1]:150-155
8. Hublin C, Kaprio J, Partinen M, Heikkilä K, Koskenvuo M (1996) Daytime sleepiness in an adult, Finnish population. J Int Med 239:417-423
9. Parkes JD (1993) Daytime sleepiness. BMJ 306:772-775
10. Stone DH (1993) Design a questionnaire. BMJ 307:1264-1266
11. Sackett DL, Haynes RB, Guyatt GH, Tugwell P (eds) (1991) Clinical epidemiology. A basic science for clinical medicine. Little, Brown and Company, Boston, Massachussetts, USA
12. Karacan I, Thornby JI, Anch M, Holzer CE, Warheit GJ, Schwab JJ, Williams RL (1976) Prevalence of sleep disturbance in a primarily urban Florida county. Soc Sci Med 10:239-244
13. Bixler EO, Kales A, Soldatos CR, Kales JD, Healey S (1979) Prevalence of sleep disorders in the Los Angeles metropolitan area. Am J Psychiatry 136:1257-1262
14. Lugaresi E, Cirignotta F, Zucconi M, Mondini S, Lenzi PL, Coccagna G (1983) Good and poor sleepers: an epidemiological survey of the San Marino population. In: Guilleminault C, Lugaresi E (eds) Sleep/wake disorders: natural history, epidemiology, and long-term evaluation. Raven Press, New York, pp 1-12
15. Klink M, Quan SF (1987) Prevalence of reported sleep disturbances in a general adult population and their relationship to obstructive airways diseases. Chest 91:540-546
16. Ford DE, Kamerow DB (1989) Epidemiologic study of sleep disturbances and psychiatric disorders. JAMA 262:1479-1484
17. Phillips B, Cook Y, Schmitt F, Berry D (1989) Sleep apnea: prevalence of risk factors in a general population. South Med J 82:1090-1092
18. Schmidt-Nowara WW, Wiggins CL, Walch JK (1991) Sleepiness in an adult population: prevalence, validity, and correlates. In: Peter JH, Penzel T, Podszus T, von Wichert P (eds) Sleep and health risk. Springer-Verlag, Berlin Heidelberg, pp 78-83
19. Zeitlhofer J, Rieder A, Kapfhammer G, Bolitschek J, Skrobal A, Holzinger B, Lechner H, Saletu B, Kunze M (1994) Zur Epidemiologie von Schlafstörungen in Österreich. Wiener Klinische Wochenschrift 106:86-88
20. Partinen M, Rimpelä M (1982) Sleeping habits and sleep disorders in a population of 2016 finnish adults. The Yearbook of Health Education Research. The National Board of Health. Finland, pp 253-260
21. Broman JE, Lundh LG, Hetta J (1996) Insufficient sleep in the general population. Neurophysiol Clin 26:30-39
22. Partinen M, Putkonen PTS (1981) Sleep habits and sleep disorders in 2537 young Finnish males. Sleep 1980. 5th Eur Congr Sleep Res, Amsterdam 1980. Karger, Basel, pp 383-385
23. Billiard M, Alperovitch A, Perot C, Jammes A (1987) Excessive daytime somnolence in young men: prevalence and contributing factors. Sleep 10:297-305
24. Gislason T, Almqvist M (1987) Somatic diseases and sleep complaints. An epidemiological study of 3201 Swedish men. Acta Med Scand 221:475-481

25. Stradling JR, Crosby JH, Payne CD (1991) Self reported snoring and daytime sleepiness in men aged 35-65 years. Thorax 46:807-810

26. Gislason T, Benediktsdóttir B, Björnsson JK, Kjartansson G, Kjeld M, Kristbjarnarson H (1993) Snoring, hypertension, and sleep apnea syndrome. An epidemiologic survey of middle-aged women. Chest 103:1147-1151

27. Ancoli-Israel S, Kripke DF, Klauber MR, Mason WJ, Fell R, Kaplan O (1991) Sleep-disordered breathing in community-dwelling elderly. Sleep 14:486-495

28. Martikainen K, Urponen H, Partinen M, Hasan J, Vuori I (1992) Daytime sleepiness: a risk factor in community life. Acta Neurol Scand 86:337-341

29. Gislason T, Reynisdóttir H, Kristbjarnarson H, Benediktsdóttir B (1993) Sleep habits and sleep disturbances among elderly - an epidemiological survey. J Int Med 234:31-39

30. Jennum P, Sjøl A (1994) Self-assessed cognitive function in snorers and sleep apneics. Eur Neurol 34:204-208

31. Foley DJ, Monjan AA, Brown SL, Simonsick EM, Wallace RB, Blazer DG (1995) Sleep complaints among elderly persons: an epidemiologic study of three communities. Sleep 18:425-432

32. Janson C, Gislason T, De Backer W, Plaschke P, Björnsson E, Hetta J, Kristbjarnason H, Vermeire P, Boman G (1995) Daytime sleepiness, snoring and gastro-oesophageal reflux amongst young adults in three European countries. J Int Med 237:277-285

33. Bliwise DL, King AC (1996) Sleepiness in clinical and nonclinical populations. Neuroepidemiology 15:161-165

34. Enright PL, Newman AB, Wahl PW, Manolio TA, Haponik EF, Boyle PJR (1996) Prevalence and correlates of snoring and observed apneas in 5,201 older adults. Sleep 19:531-538

35. Ganguli M, Reynolds CF, Gilby JE (1996) Prevalence and persistence of sleep complaints in a rural older community sample: the MoVIES Project. J Am Geriatr Soc 44:778-784

36. Hays JC, Blazer DG, Foley DJ (1996) Risk of napping: excessive daytime sleepiness and mortality in an older community population. J Am Geriatr Soc 44:693-698

37. D'Alessandro R, Rinaldi R, Cristina E, Gamberini G, Lugaresi E (1995) Prevalence of excessive daytime sleepiness. An open epidemiological problem. Sleep 18:389-391

38. Johns MW (1991) A new method for measuring daytime sleepiness: the Epworth Sleepiness Scale. Sleep 14:540-545

39. Buysse DJ, Reynolds III CF, Monk TH, Berman SR, Kupfer DJ (1989) The Pittsburg Sleep Quality Index: a new instrument for psychiatric practical and research. Psychiatry Res 28:193-213

40. Rosenthal L, Roehrs TA, Roth T (1993) The Sleep-Wake Activity Inventory: a self-report measure of daytime sleepiness. Biol Psychiatry 34:810-820

41. Cornoni-Huntley J, Ostfeld AM, Taylor JO, Wallace RB, Blazer D, Berkman LF, Evans DA, Kohout FJ, Lemke JH, Scherr PA, Korper SP (1993) Established populations for epidemiologic studies of the elderly: study design and methodology. Aging Clin Exp Res 5:27-37

42. Rinaldi R, D'Alessandro R, Sforza E, Bassein L, Cristina E, Gamberini G, Lugaresi E (1997) Validation of a new questionnaire investigating symptoms related to excessive daytime sleepiness (submitted)

43. Schoenberg BS (1983) Calculating confidence intervals for rates and ratios. Neuroepidemiology 2:257-265

Breathing Disorders and Sleep

G. COCCAGNA AND F. CIRIGNOTTA

Sleep Apnea Syndrome

In the 1950s some American authors identified a syndrome of chronic alveolar hypoventilation linked to severe obesity. Burwell et al. [1] coined the term "Pickwickian syndrome" after the character of fat Joe created by Charles Dickens in the *Posthumous Papers of the Pickwick Club*. This character presents all the clinical features of the syndrome he epitomizes: severe obesity, marked daytime somnolence, voracious appetite, noisy breathing during sleep and, as the colour of his face suggests, polycythemia.

Back in the 1950s, pneumologists accepted the hypothesis postulated by Burwell et al. that the syndrome was caused by an increased respiratory workload due to obesity with a consequent rise in O_2 consumption and CO_2 production. The syndrome was exacerbated by the body's progressive adjustment to the condition of alveolar hypoventilation.

In the 1960s, interest in Pickwickian syndrome shifted from pneumology to sleep medicine. Sleep laboratories in Germany [2], France [3] and Italy [4] demonstrated that the syndrome is caused by obstructive apneas occurring as soon as the patient falls asleep and persists throughout the night. In one Italian sleep laboratory, Coccagna et al. [5] applied sophisticated polysomnographic recording techniques and highlighted the severe repercussions such apneas had on ventilatory and cardiocirculatory functions. As meanwhile it was discovered that a clinico-polygraphic picture identical to Pickwickian syndrome may be found in non-obese subjects, hypnologists decided to group all these syndromes under the heading "obstructive sleep apnea syndromes" (OSAS). The Bologna team were also the first to link the syndrome with snoring for the simple reason that subjects with OSAS had previously been heavy snorers for years or even decades [6]. It was then demonstrated that trivial snoring is the first step of a process which may gradually evolve into the complicated form of OSAS. Lugaresi et al. [7] then proposed the term "heavy snorers' disease" to classify the sum of clinical features caused by upper airway obstruction during sleep.

Istituto di Neurologia, Università di Bologna, Via Ugo Foscolo 7, 40123 Bologna, Italy

Polygraphic Aspects of Upper Airway Obstruction During Sleep

Snoring

Snoring is a mainly inspiratory noise caused by vibration of the soft palate and posterior faucial pillars. Polysomnographic recordings have demonstrated that it is linked to sleep-related stenosis of the upper airways, as indicated by the increased negative endothoracic pressure and elevated electromyographic activity of intercostal muscles during the noisy inspiratory acts. Snoring peaks in intensity during deep slow sleep while obstructive apneas, alone or in short clusters may appear in light and REM sleep in heavy snorers. During snoring mild alveolar hypoventilation is present; systemic arterial pressure does not fall as it does in normal subjects, but remains unchanged or slightly elevated.

Obstructive Sleep Apnea Syndrome

When obstructive apneas become numerous or continuous, the patient is said to have OSAS and snoring during the night becomes intermittent. In our opinion the commonly accepted limit of five apneas per hour (apnea index = A.I. $\geq$ 5) is too low to establish the dividing line between health and disease as it is close to the physiological values encountered in the elderly. An A.I. $\geq$ 10 seems to be a more realistic boundary line.

Three types of apnea have been identified by polygraphic recordings:

1. *Central apneas* are said to occur when chest movements stop and resume simultaneously with airflow through the upper airways. These apneas are the fewest (2.3%) and shortest (10-20 s), and occur almost exclusively during light sleep.
2. *Obstructive apneas* are characterized by an arrest of airflow through the upper airways while chest movements persist and progressively intensify until breathing is resumed. The chest movements during the apnea indicate the inspiratory effort to overcome the upper airway obstruction as displayed by concomitant recordings of endoesophageal pressure which demonstrate wide negative oscillations during inspiration. These apneas are the most numerous (over 85%) and the longest (even over 3 min).
3. *Mixed apneas* start with a complete breathing arrest lasting several seconds (central apnea), followed by the appearance of chest movements while airflow through the upper airways remains blocked (obstructive apnea). These apneas are frequent (12.5%), prevalent in REM sleep and may be prolonged.

Apneas cease with a brief arousal. Since upper airway patency is never complete, 4-8 noisy breathing acts preceed the next apnea. Intermittent snoring is therefore a key diagnostic marker for OSAS.

The Ventilatory and Cardiocirculatory Effects of Apneas

Polygraphic recordings during sleep monitoring systemic and pulmonary arterial pressure, ECG, ear oximetry, and regular arterial blood sampling have established that major hemodynamic and ventilatory changes take place during obstructive or mixed apneas [5, 8].

a) During apneas there is a state of alveolar hypoventilation which gradually intensifies in the subsequent stages of slow sleep, becoming severely exacerbated in REM sleep when the apneas are most prolonged. We observed anoxic-asphytic attacks in three patients triggered by prolonged apneas during REM sleep episodes. The tonic seizure due to anoxia interrupted the apnea 4-5 min after the arrest of breathing [9].

Ear oximetry recording showed that until the complicated stage of the disease is reached, the breathing acts occurring after each apnea are sufficient to restore O_2 saturation to normal levels.

b) Alongside alveolar hypoventilation, pulmonary hypertension develops and peaks in REM sleep. Pressure is highest at the end of each apnea, when breathing is resumed.

c) Systolic and diastolic systemic arterial pressure during the apnea tend to rise gradually with a parallel increase in heart rate. Alternatively, the rise in arterial pressure may be accompanied by progressive bradycardia. Less frequently, arterial pressure and heart rate remain unchanged during the apnea; more rarely still a gradual fall in pressure may occur accompanied by brady- or tachycardia.

The variable behaviour of arterial pressure during obstructive apneas is due to the complex array of biochemical, mechanical and neurovegetative events which occur during the apnea and often interfere antagonistically in different ways from patient to patient [10].

As ventilation is resumed, there is always a sudden sharp rise in arterial pressure with an increase in pulse pressure. At the end of the apnea, pressure values are always much higher than awake values and peak in REM sleep.

d) Manifold cardiac arrhythmias may apppear during apneas, especially bradyarrhythmias, the most frequent being sinus bradycardia. In one of our patients, heart blocks lasting 8-13 s occurred with almost every apnea during REM sleep. Other arrhythmias such as first and second degree A-V blocks and ventricular extrasystoles have also been observed. Cardiac arrhythmias are probably due to two factors: strong vagal stimulation and marked hypoxia during the apnea.

Pathophysiology of Upper Airway Obstruction During Sleep

A number of anatomical factors such as obesity, adenoid or tonsillar hypertrophy, micrognathia, macroglossia and acromegaly restrict the upper airways and foster the onset of obstructive apneas. However, not everyone with these abnormalities has nocturnal apneas so that other factors must intervene. One additional cause of upper airway obstruction could lie in the failure of the dilator muscles in the pharynx (particularly the genioglossus) to synchronize with the diaphragm: when pharyngeal muscle contraction is weak or delayed, the lumen of the pharynx is narrowed when the diaphragm contracts. Collapse of the oropharyngeal walls is also favoured by the supine position and by sleep-induced physiological muscle hypotonia. In fact, when hypotonia is exacerbated by sleep deprivation, alcohol use or benzodiazepine intake, there is an increase in the number and length of apneas. A recent report demonstrated that heavy snorers have an anomaly in muscle fibre distribution in the pharyngeal dilator muscles: these patients have a smaller proportion of type I and type IIb fibres and a higher percentage of type IIa fibres than do controls. The type IIa fibres are also hypertrophic [11].

Controversy surrounds the role of a central factor in triggering apneas. However, the fact that as snoring gradually progresses to OSAS the apneas appear first in light sleep and REM sleep, suggests that unstable breathing control upon falling asleep and reduced excitability of the respiratory centre in REM sleep must be implicated in the onset of apneas. Finally, progressively attenuated responses to chemical stimuli on the part of the respiratory centre could underlie the autoexacerbating tendency of the syndrome to evolve into the complicated form of OSAS (see below).

Apneas probably cease as a result of two mechanisms, one mechanical, the other chemical. Both trigger an arousal of varying duration and consequent activation of the pharyngeal dilator muscles. The mechanical stimulus consists of a proprioceptive reflex from the upper airways which house the receptors sensitive to pressure and stretching. During the apnea these receptors receive increasing stimulation until they trigger an arousal through some unknown path. The chemical stimulus is the hypoxia which gradually increases during the apneas and activates the carotid chemoreceptors leading to an arousal.

The longer duration of apneas in REM sleep is therefore due to two factors: the reduced sensitivity of the respiratory centres to hypoxia and the reduced intensity of the proprioceptor stimuli located in the upper airways caused by the REM sleep-linked weakness of the inspiratory muscles struggling to overcome the obstruction.

Natural History of OSAS

It is believed that there is a clinical continuum from snoring to the full-blown or complicated form of OSAS. For this reason we proposed dividing heavy snorers' disease into four evolutive stages according to clinical and polygraphic criteria [7]:
- Stage 0 (*preclinical*). Corresponds to continuous or almost all-night heavy snoring. The intensity and continuity of snoring may depend on sleeping position (back or side), smoking, alcohol intake, or sleep deprivation. Sporadic obstructive apneas may occur in light and REM sleep.
- Stage 1 (*initial*). Obstructive apneas recur virtually without a break in light sleep (stages 1-2) and REM. Snoring therefore becomes intermittent. In slow deep sleep snoring is continuous as apneas are absent. Daytime drowsiness may appear during this stage.

- Stage 2 (*full-blown*). Obstructive apneas recur uninterruptedly throughout sleep in all sleep stages (Fig.1). Snoring is always intermittent. Severe daytime somnolence is virtually constant. Signs of cardiocirculatory impairment are present (right heart overload at ECG, hypertension, hepatomegaly, etc.).

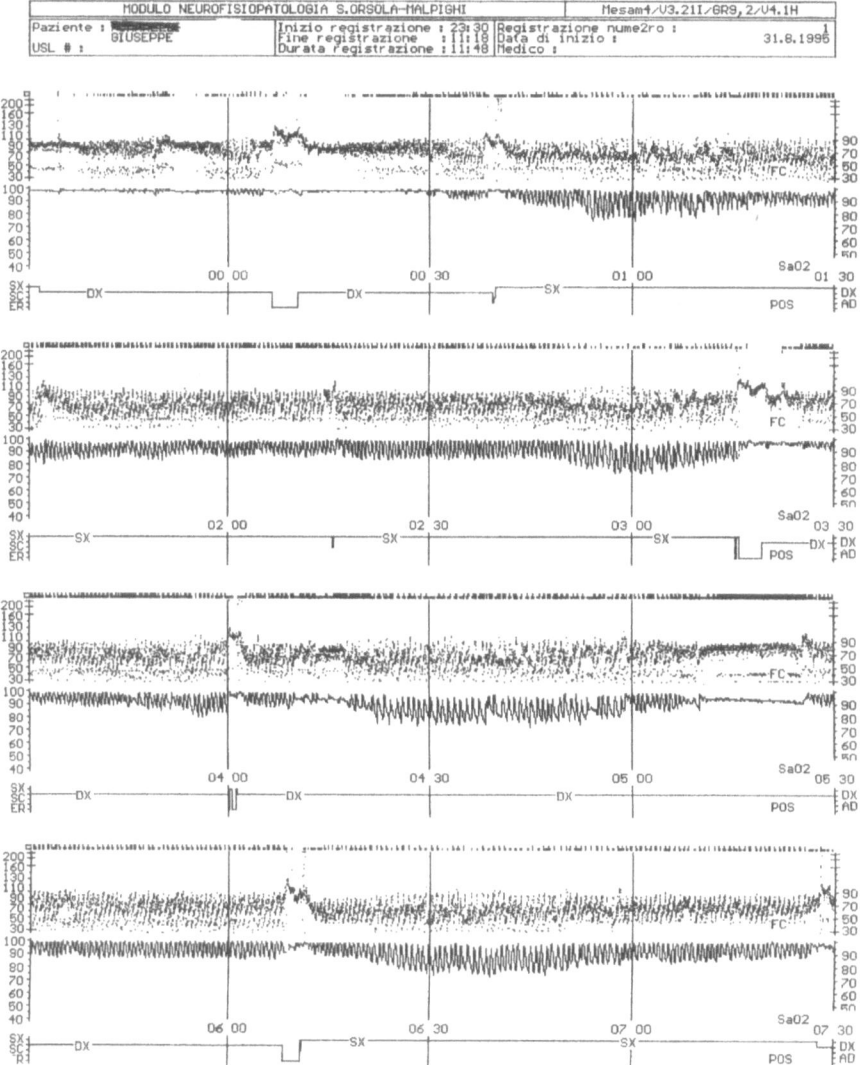

Fig.1. *Nocturnal ambulatory polysomnographic recording (Mesam IV, MAP, Martinsried, Germany).* Parameters (*from top to bottom*): breathing sound, heart rate, oxygen saturation, sleeping position (*back or side*). Patient with OSAS. A continuous series of O_2 desaturation events are accompanied by intermittent snoring and heart rate oscillations. The trend of these parameters will establish a diagnosis of OSAS

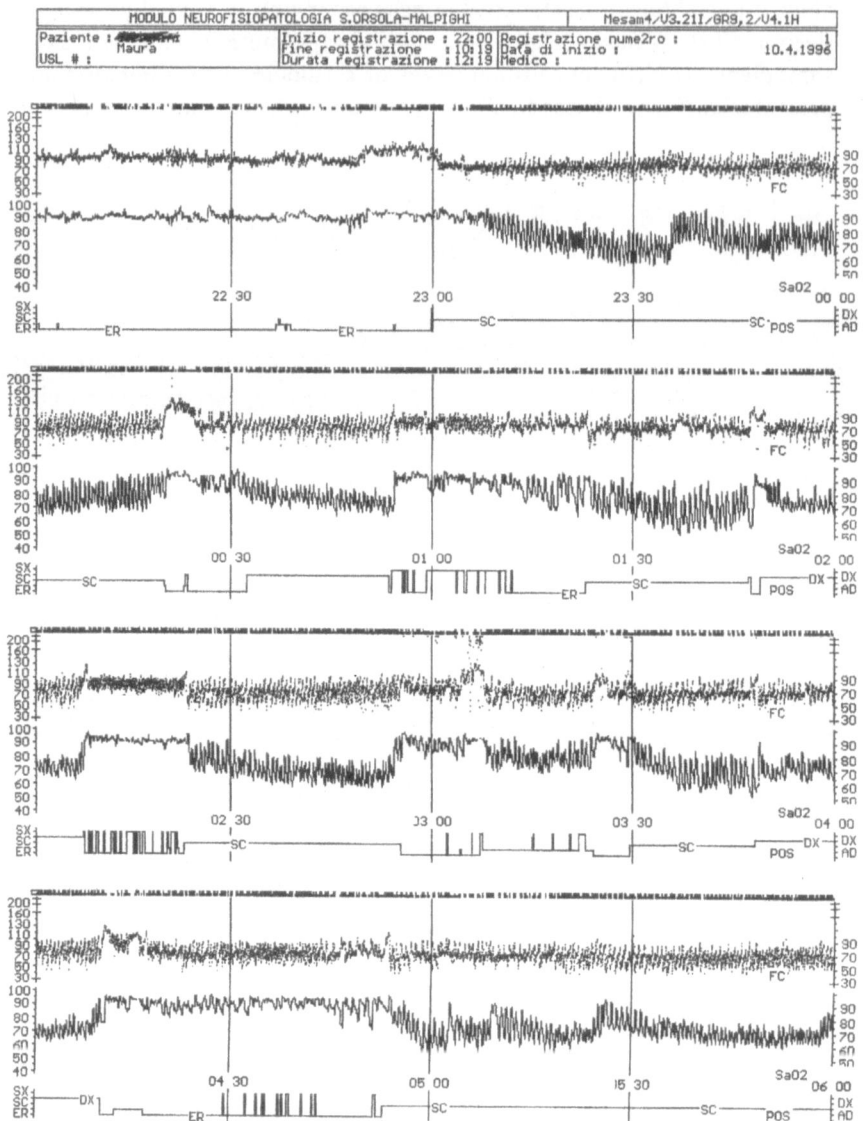

Fig. 2. *Nocturnal ambulatory polysomnographic recording (Mesam IV, MAP, Martinsried, Germany).* Parameters *(from top to bottom)*: breathing sound, heart rate, oxygen saturation, sleeping position *(back or side)*. Patient with Overlap Syndrome. The O_2 desaturation events linked to the apneas overlap a progressive fall in baseline SaO_2 linked to COPD

- Stage 3 (*complicated*). Apneas recur throughout sleep, but unlike what happens in stage 2, SaO_2 fails to return to normal values when ventilation is resumed at the end of each apnea. This is particularly apparent in REM sleep: ear oximetry had demonstrated that SaO_2 progressively declined during REM (Fig. 2). At this stage the disease is severe, daytime somnolence is remarkable, alveolar hypoventilation persists during wakefulness alongside polycythemia and cyanosis, and there are clear signs of right heart failure (peripheral edema, hepatomegaly, retinic vein congestion and sometimes papilledema, cardiomegaly). Over 50% of patients present stable systemic hypertension.

Exactly how the syndrome develops into the complicated stage remains a matter of debate. Some claim that this occurs when the classical features of OSAS are accompanied by a chronic obstructive pulmonary disease - a frequent but by no means constant association. Another hypothesis is that the respiratory centres and carotid chemoreceptors become progressively less sensitive to blood gases leading to ventilatory failure: the syndrome progressively worsens as the body struggles to adapt to increasingly elevated CO_2 values.

Patients with complicated OSAS do have a much lower ventilatory response to CO_2 than subjects without alveolar hypoventilation during wakefulness. Whether or not this reduced CO_2 response is acquired or congenital remains unknown.

Clinical Features of OSAS

Age at onset of clinical symptoms often depends on the etiology. When hypertrophied tonsils and/or adenoids are implicated, the syndrome usually appears before the age of eight. In the pickwickian syndrome, the full-blown disease usually develops over the age of 40, after the patient has been a heavy snorer since the second or third decade of life. In other upper airway malformations (e.g. micrognathia), the syndrome may arise many years after the appearance of the anatomical alteration. OSAS is prevalent in men (82% of all cases); in women the disease usually appears after menopause. Hormonal factors clearly play a major role both because they are responsible for the different shape of the pharynx and larynx in the two sexes and because some hormones, like progesterone, stimulate the respiratory centre.

Clinically, the disease is held to be established when two basic symptoms are present: intermittent snoring and daytime somnolence. Daytime somnolence is constant and severe in stage III or complicated heavy snorer's disease, but can also be found in patients with relatively few apneas. Sometimes a patient may complain of insomnia due to the recurrent awakenings throughout the night. Although the patient is often unaware of it, nocturnal sleep in OSAS is always severely disrupted due to the marked loss of deep slow sleep. Morning headache and stupor are commonly noted by the patient along with nocturia and nocturnal enuresis, the latter especially frequent in children.

Sleep in OSAS patients is described as agitated with disordered and uncoordinated movements at the end of each apnea, sometimes making patients fall out of bed. Nocturnal myoclonus may be present. Behavioural disturbances such as aggressiveness and irritability are often reported by patients alongside impaired cognitive function. The patients described by Kales et al. [12] tested by the Minnesota Multiphasic Personality Inventory (MMPI) presented the typical profile of the somatic – neurotic type with high scores for depression (56%), hypochondriasis (35%) and conversion-hysteria (29%); 76% of patients had "suspected or mild to severe deficits in terms of thinking, perception, memory, communication, or the ability to learn new information". Patients also referred to a frequent and severe psychosocial disruption in their lives.

Roughly 30% of patients present sexual problems varying from reduced libido to impotence. Some metabolic and hormonal changes are frequent. GH, testosterone and LH secretion are diminished during sleep as are plasma levels of aldosterone, cortisol and plasma renin activity. Instead, excretion of catecholamines, atrial natriuretic factor and TSH hormone is increased.

Nocturnal Upper Airway Obstruction and Cardiocirculatory Diseases

Arterial Hypertension

The first epidemiological survey to investigate a possible correlation between snoring and stable hypertension was carried out by the Bologna team in the Republic of San Marino [13]. From that survey it appeared that the trend of systemic arterial pressure values in heavy snorers was significantly higher than the blood pressure values of controls, irrespective of body weight. The relation was particularly significant in the age group 41-60 years, where hypertension was present in 15.2% of habitual snorers and in 7.5% of non-snorers. A similar prevalence of hypertension in heavy snorers was confirmed by subsequent epidemiological studies [14-17]. Schmidt-Nowara et al. [17] found a higher incidence of hypertension and heart disease among snorers. After adjustment for confounding factors, this correlation proved significant for myocardial infarction but not for hypertension. Telakivi [18] found that only the body mass index was an independent determinant of diastolic blood pressure in heavy snorers. Recently, Hla et al. [19] in a cross-sectional study of blood pressure during wakefulness and sleep in an adult population with and without sleep-related disordered breathing found a significant association between hypertension during both wakefulness and sleep and sleep apnea independent of obesity, age and sex.

The mechanisms responsible for chronic hypertension in nocturnal upper airway obstruction are not fully known. It is likely, however, that there is an increased production of catecholamines due to heightened sympathetic activity during sleep in response to the hypoxia and mechanical stress of inspiratory efforts struggling against the obstructed airways. However, other causes cannot be ruled out since the continuous haemodynamic changes accompanying the apneas, above all the abrupt pressure rises which occur when ventilation is resumed, may have important cardiovascular effects per se [20].

A number of epidemiological surveys have supplied indirect evidence of the key role played by nocturnal apneas in the development of persistent hypertension. Among patients with a diagnosis of essential hypertension, between 20% and 50% had previously undetected OSAS. These percentages were far higher than those demonstrated in control groups [21-24]. The varying percentages found by different authors probably depend on the different pressure levels taken as the cut-off limit beyond which patients were considered hypertensive and on the different A.I. used to classify OSAS patients.

In the Lavie series, for example, the prevalence of OSAS among hypertensives was 26% if an A.I. equal to or over 5 is considered, and 22% if the A.I. limit is 10 or more. However, the most clear-cut proof that obstructive apneas are directly responsible for persistent hypertension in wakefulness is the fact that pressure values normalize when the apneas subside as well as after tracheostomy [25], weight loss [21] and after treatment with continuous positive airway pressure [26].

Heart Disease

Several epidemiological studies have stressed the association between snoring and ischemic heart disease [17, 27, 29]. On the other hand, obstructive disordered breathing during sleep has commonly been found in patients with coronary heart disease [30].

In Hung et al.'s [27] large case-control study not only was there a significant association between sleep apnea and myocardial infarction, but also an increased risk of myocardial infarction with increasing levels of sleep apnea. The association was independent of age, body mass index, arterial hypertension, smoking and cholesterol level.

In a Bologna case-control study heavy snorers proved to have a significantly higher incidence of myocardial infarction than a control population. Multivariate analysis included hypertension diabetes, smoking and alcohol abuse [31].

Brain Infarction

Habitual snorers had a significantly higher risk of brain infarction than occasional or never snorers according to a case-control study of 50 male stroke patients and 100 male controls [32]. Palomaki et al. [33] found a significant correlation between heavy snoring and the onset of brain infarction during the night or immediately after awakening in 177 consecutive male patients with brain infarction. In a multiple logistic regression analysis the risk remained elevated even when age, arterial hypertension, body mass index, smoking, alcohol consumption and diabetes mellitus were tested. Snoring was not a risk factor in patients who presented brain infarction during the day. Arterial hypertension seemed to have an additive effect on the independent risk caused by snoring.

Is Snoring per se a Risk Factor for Cardiovascular Disease?

Most surveys on the correlation between snoring and cardiovascular disease are based on questionnaires. To make results more significant and uniform, only those subjects who declare themselves to be every night or regular snorers, i.e. heavy snorers, are usually taken into consideration. But how uniform is this population? To solve this problem our team made a survey on 3479 30- to 69-year-old men living in Bologna [34]. Ten per cent of those answering the questionnaire declared themselves to be "every night snorers". Most were aged between 50 and 60. We invited 40 randomly selected "every night snorers" to undergo nocturnal polygraphic recording. The results showed than 14 men (35%) had an apnea index over 10; two of them had daytime somnolence and one was already in stage II of the full-blown disease.

In conclusion, our findings showed that epidemiological data on snoring obtained from questionnaries are relatively unreliable as they do not discriminate between the initial or full-blown forms of OSAS. As Waller and Bhopal [35] emphasized, we cannot claim with certainty that trivial snoring per se has critical haemodynamic effects.

Hoffstein et al. [36] monitored 372 snorers overnight and found that diastolic blood pressure correlated significantly with body mass index, apnea-hypopnea index and mean oxygen saturation, but not with snoring index (number of snores per hour of sleep). As the snoring index correlated with body mass index, apnea-hypopnea index and mean nocturnal oxygen saturation, the authors concluded that snoring is not a direct risk factor for hypertension but may influence blood pressure via its association with the above mentioned variables. More recently, Hoffstein [37] pointed out that it will only be possible to establish whether snoring is dangerous to health by tackling the problem differently. Longitudinal polysomnographic cohort studies are required on groups of snorers once the parameters defining snoring have been defined and standardized.

Hence the problem of when snoring ceases to be an irksome noise and becomes a disease in its own right remains open.

Epidemiology of Snoring and OSAS

The first epidemiological survey on snoring done in the Republic of San Marino (northeast Italy) demonstrated that about 24% of men and 14% of women snore habitually [13]. The number of snorers rises with age so that 60% of men and 40% of women aged between 60 and 65 are regular snorers. After this age the number of snorers decreases [34]. Schmidt-Nowara et al. [17] found similar percentages in a Hispanic-American population in which the age-adjusted prevalence of regular loud snoring was 27.8% for men and 15.3% for women. Among Finns, there were fewer regular snorers (9% of adult men and 3.6% of adult women) [28]. Billiard et al. [38] reported 13.6% habitual snoring in a population of young men aged 17-22. Among these subjects, 31.9% complained of daytime sleep episodes.

The prevalence of OSAS varies in different epidemiological studies and depends on the different criteria adopted in selecting the study population. The highest prevalence of OSAS was found in Bologna [34] with 2.5% of the male population aged 30-69 years presenting heavy snorers' disease (stage 1 or higher). The age distribution curve for OSAS was similar to that of snoring. Other studies report a much lower prevalence of OSAS.

Lavie [39] found that the prevalence of OSAS among industrial workers in Israel was around 1%, while in Finland the estimated prevalence of OSAS among 40- to 50-year-old men was between 0.4% and 1.4% [40]. In a study on 30- to 69-year-old Swedish men, Gislason [41] found a 1.3% prevalence ratio of the syndrome, but their survey included a number of occasional snorers.

Treatment

OSAS treatment is based on weight loss, treatment with nasal positive airway pressure, surgery, orthodontic appliances and other techniques. The choice of therapy depends on an overall assessment of numerous factors: severity of the nocturnal breathing disorder (number of apneas, degree of O_2 desaturation, cardiac arrhythmias), severity of daytime somnolence, patient characteristics (age, constitution, anatomical abnormalities of the airways, occupation, personality), the social impact of the disease and concomitant risk factors (heart disease, hypertension). All these elements are pooled and assessed in relation to the indications and limitations of the therapeutic strategies available.

Weight Loss

OSAS patients often report the onset or exacerbation of the breathing disorder coinciding with weight gain. Although weight loss was among the first treatments proposed for the syndrome, literature reports on the topic are scant: the case series are small and breathing indices often incomplete. In addition, cohort studies on the long-term effects are lacking [42, 43]. The major criticism of slimming treatments concerns the difficulty in achieving and maintaining adeguate weight loss in the long term. Our group studied 35 patients who underwent nocturnal polygraphic recordings before and after weight loss achieved by a low calorie or semi-liquid diet over a period of eight months [44]. As a result of the diet, average BMI (body mass index) fell from 36.2 to 29.6 and the average apnea index dropped from 60 to 30. However, further detailed analysis of the series showed that only 34% of patients could be deemed clinically cured (i.e. with an apnea index below 10), while the remaining 66%, albeit improved to different degrees, continued to present an apnea index over 10 and thus required further treatment. These results were independent of the amount of weight lost. Negative prognostic factors were the severity of the syndrome under basal conditions and the presence of anatomical airway abnormalities.

Nasal Positive Pressure

Nasal continuous positive airway pressure (CPAP), introduced by Sullivan et al. in 1981 [45] is based on the principle of creating in the upper airways a positive pressure a few centimetres of water above atmospheric pressure, usually not more than 15 cm H_2O, which will keep the airways patent during inspiration. The pressure is generated by an air pump that blows room air into the patient's oropharyngeal walls through a nasal mask attached to a flexible tube. The correct pressure for each patient is determined by nocturnal polysomnographic recordings during which pressure is progressively increased until the apneas and then snoring disappear. The pressure required to make the apneas cease varies in relation to the patient's position and sleep stage. Currently prescribed pressure is that required to suppress apneas and snoring in the worst situation, i.e. during REM sleep in the supine position. Bearing in mind that patients will require a lower pressure for most of the night, new "autotitrating" equipment has been devised to adjust the pressure erogated throughout the night.

CPAP may be poorly tolerated by patients because of mask discomfort and nasal dryness and irritation, or when high pressure values are required. To overcome the latter drawback, a new piece of equipment has been introduced which produces two different pressure values for inspiration and expiration, respectively. This device, the bilevel positive airway pressure system (BiPAP) is more expensive, but better tolerated by patients who cannot use CPAP.

CPAP and BiPAP are symptomatic treatments: symptoms reappear as soon as the patient stops using the device. In a few cases long-term use may have a curative effect, eliminating self-aggravating anatomical and functional factors along with the apneas.

Surgery

Surgery is indicated when OSAS is due to specific anatomical abnormalities which can be corrected surgically, or when other noninvasive treatments fail or are poorly tolerated. It is important to inform patients about the risk-benefit ratio of surgery and the existence of other treatments [46].

The surgical procedures most widely performed [47, 48] include operations on the turbinates and nasal septum, tonsillectomy, uvulopalatopharyngoplasty (UPPP) and uvulopalatal flap, glossectomy, lingual plastic surgery, inferior sagittal mandibular osteotomy with myotomy and hyoid bone suspension, osteotomy and maxillomandibular advancement, and tracheostomy.

Surgery on nasal structures proves effective in a low percentage of cases and only when symptoms are mild, but may improve CPAP use in cases of nasal stenosis. Tonsillectomy is mainly performed in children and can resolve even severe cases. Uvulopalatopharyngoplasty is a procedure which widens the retropalatal spaces by incision and advancement of the palatine pillars, cutting the uvula and removing the tonsils. The operation is indicated in mild cases with an apnea index under 20 and a BMI under 28. Surgical candidates should be rigorously selected, undergoing endoscopy to determine the level of obstruction by simulating an apnea with a Muller manoeuvre. The obstruction may be retropalatal or retrolingual: in retrolingual obstructions, surgery is unlikely to relieve symptoms. UPPP entails specific risks such as pain, bleeding and dysphagia. In addition, excessive resection of palatal tissue may lead to velopharyngeal incompetence with vocal changes and nasal regurgitation. To overcome this risk a new, less destructive surgical technique, the uvulopalatal flap, has been devised by which the uvular tip is repositioned toward the hard palate. In the case of retrolingual obstruction, glossectomy and linguoplasty procedures have been performed with limited resection of the lateral and dorsal part of the tongue. Maxillofacial surgery is indicated when cephalometry discloses a reduction of the posterior air space with the base of the tongue very close to the posterior wall of the pharynx. The simplest procedure entails osteotomy and advancement of the anterior mandible which includes the attachment of the genioglossus muscle with hyoid myotomy and suspension. The more complex procedures involve maxillomandibular osteotomy with advancement of bony structures.

Tracheostomy resolves the syndrome completely. A fenestrated tracheal cannula is inserted which is kept closed by day and opened at night. However, local inflammation and the ensuing tracheal stenosis and psychological repercussions have limited this operation to emergency cases only, when the severity of apneas is life-threatening and CPAP cannot be used.

Orthodontic Appliances and Other Techniques

Over the years numerous other treatments have been put forward, many of which are minor and often only curiosities. Few publications have assessed the effectiveness of prosthetic appliances which force the jaw or tongue forwards during sleep. These devices are not without side effects affecting the temporomandibular joint and are often poorly tolerated by patients [49].

Starting from the fact that the supine position favours the appearance of apneas, some attempts have been made to force patients to sleep on their sides by sewing a tennis ball into the back of their pyjamas.

A plaster has recently been marketed to be applied to the sides of the nose. By means of a semirigid plastic centrepiece the plaster slightly dilates the nostrils reducing nasal resistance. The plaster is a small contribution to interrupting the pathophysiological chain of events which lead to snoring, but is not effective against apneas. It has yet to be demonstrated whether the plaster is an effective measure to prevent snoring.

Medical Management

There are no drugs available to treat OSAS. Respiratory analeptics, progesterone, tricyclic drugs, naloxone, propranolol and tryptophan have proved ineffective or have given modest short-term benefit. O_2 administration during sleep generally raises SaO_2 but also leads to a concomitant rise in the length of apneas with hypercapnia and respiratory acidosis. O_2 administration is therefore not recommended, especially in patients who are already hypercapnic during wakefulness.

Alcohol intake in the evening may exacerbate the apneas. Anxiolytic and hypnotic drugs may also favour apneas by reducing muscle tone and depressing the respiratory centres. However, literature data on hypnotics do not confirm theoretical premises and further studies are necessary to clarify the problem [50].

Central Alveolar Hypoventilation Syndrome

In the 1950s a chronic hypoventilation syndrome was first described in the absence of obesity, chronic obstructive pulmonary disease or musculoskeletal abnormalities of the breathing apparatus. The syndrome was ascribed to a primary impairment of the respiratory centres and was therefore termed central, primary, idiopathic or neurogenic alveolar hypoventilation. In 1962, observing the syndrome in three patients who had undergone surgery to the upper cervical spine, Severinghaus and Mitchell [51] coined the expression "Ondine's curse" after the hero of Giraudoux's play [52]. Guilty of having betrayed the love of Ondine the nymph, Hans the knight was punished by being deprived of all the body's automatic functions including breathing, which could only be carried out as a voluntary act; he only had to fall asleep to stop breathing. By referring to the fate of Ondine's lover, Severinghaus and Mitchell wanted to emphasize how peculiar this syndrome was: a condition which arose and was aggravated during sleep. There exist idiopathic forms which are almost always congenital and may be accompanied by malformations of the neural crest (multiple ganglioblastomas, ganglioneuroma, Hirschprung's disease).

Patients with central alveolar hypoventilation present cyanosis during sleep, headache on awakening, daytime somnolence and above all, episodes of acute respiratory failure with impaired consciousness which may even turn into coma. These episodes mainly occur during sleep and coincide with even mild airway disorders, administration of sedatives, general anaesthesia, etc. Patients require hospital admission and assisted ventilation. Respiratory centre stimulation with CO_2 always shows a marked fall in ventilatory response. During wakefulness arterial blood gases may be normal or show only slightly impaired values, but a short voluntary hyperpnea promptly restores O_2 and CO_2 values to within normal limits. In adults central alveolar hypoventilation is usually secondary to a variety of brain stem lesions (poliomyelitis, stroke, tumours, surgery, etc.); the condition is occasionally reversible.

Polysomnographic recordings have demonstrated that during sleep alveolar hypoventilation develops accompanied by a fall in the amplitude of rib cage and abdominal excursions. Mainly central apneas may occur, but do not coincide with the lowest SaO_2 values. In congenital or infantile forms of the syndrome, the alveolar hypoventilation is more pronounced in deep slow sleep; in adult forms, it is more pronounced in REM sleep. As far as treatment is concerned, encouraging results have been obtained using intermittent positive pressure ventilation (IPPV) during sleep with the same mask as applied to OSAS patients.

Chronic Obstructive Pulmonary Disease

Patients with chronic obstructive pulmonary disease (COPD) are usually divided into two groups, the "pink puffers" and the "blue bloaters". The former are very short of breath during wakefulness, but are normocapnic and slightly hypoxemic; the latter are less dyspnoeic, but are hypercapnic, hypoxemic and polycythemic and also have an impaired ventilatory response to CO_2. The ventilatory and haemodynamic changes during sleep described below are commonly found in blue bloaters, but are less frequent and more mild in the pink puffers.

In 1978 Coccagna and Lugaresi [53] reported the first polysomnographic study on a group of COPD patients (blue bloaters) in whom pulmonary and systemic arterial pressures were recorded throughout sleep in addition to gasanalytic values. The results showed a progressive mild increase in alveolar hypoventilation in the succesive stages of slow sleep, while a further, more significant rise occurred in REM sleep. Parallel pulmonary arterial changes occurred with a slight progressive rise in slow sleep and a sharp increase in REM (Fig. 3). On the contrary, systemic arterial pressure behaved normally, progressively dropping in slow sleep and rising again in REM. Subsequent studies using continuous SaO_2 recording during sleep by ear oximetry confirmed that peaks of severe O_2 desaturation lasting an average of 30 min would recur during REM sleep. In slow sleep these peaks were much more consistent and shorter [54]. The severity of hypoxemia reached in sleep is usually proportional to the degree of daytime hypoxemia.

Different factors combine to worsen hypoventilation during sleep, especially REM sleep: an impaired ventilation/perfusion matching, a pathological deterioration of the physiological hypnogenic hypoventilation due to diminished sensitivity of the respiratory centre to CO_2, and a reduced efficiency of the intercostal and accessory muscles of respiration.

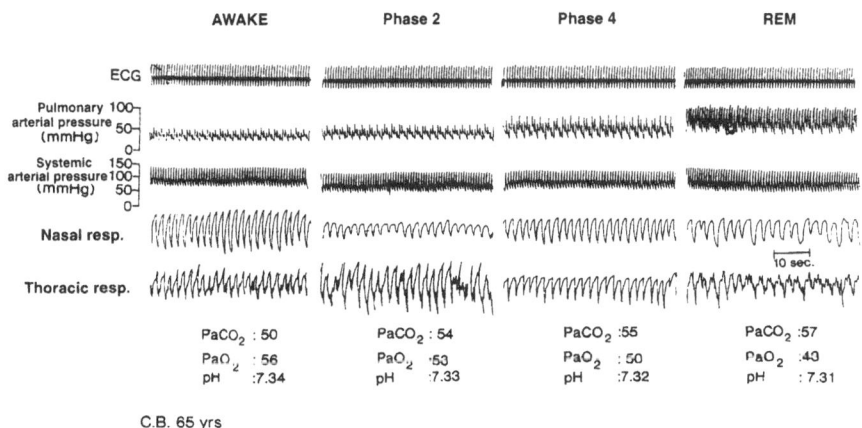

Fig. 3. *Patient with COPD.* There is a progressive increase of alveolar hypoventilation during slow sleep and a parallel rise in pulmonary arterial pressure. A further sharp rise in pulmonary pressure and alveolar hypoventilation occurs in REM sleep

Nocturnal hypoxemia can be managed by long-term oxygen therapy (administration of O_2 at a dose of 2 l/min through nasal cannulas). Intermittent positive pressure ventilation (IPPV) has also been successfully administered via a nasal mask, possibly in association with additional O_2.

COPD Combined with OSAS (Overlap Syndrome)

Some OSAS patients (10-15%) also present COPD and this prevalence is greater than that of COPD in the general population (Fig. 2). Breathing function tests should therefore be carried out in all patients with OSAS since at least some of the complicated forms of OSAS are due to this overlap. Likewise, patients with a diagnosis of COPD should undergo polysomnography when they present clinical signs of obstructive sleep apnea (snoring, especially when intermittent, daytime somnolence) or when respiratory failure is accompanied by disproportionate symptoms (e.g., polycythemia, cor pulmonale) with respect to the relatively mild degree of daytime hypoxemia. CPAP (or BiPAP) is clearly the therapy of choice in overlap syndrome to which supplemental oxygen may need to be added.

Sleep-Disordered Breathing in Neuromuscular and Skeletal Diseases

Neuromuscular disease and the rib cage deformities which often ensue may impair respiratory function to the point of alveolar hypoventilation. Hypoventilation may appear or be aggravated during sleep, particularly REM sleep, when the intercostal and accessory muscles of respiration are atonic like the other postural muscles. If the diaphragm is also involved in the neuromuscular disease, breathing during sleep may be severely impaired. Polysomnographic recordings have been performed in myasthenia, in different myopathies (Duchenne's dystrophy, nemaline myopathy, acid maltase deficiency, Steinert's myotonic dystrophy, limb girdle myopathy), motor neuron disease, kyphoscoliosis and rigid spine syndrome, but very few studies have addressed each of these diseases.

The degree of hypoventilation during sleep varies widely and is correlated more with the extent of the impairment of respiratory function tests rather than the severity of musculoskeletal deformity. None of these neuromuscular disorders presents a specific respiratory pattern during sleep, but they share two impaired breathing modalities [55]: (a) episodes of O_2 desaturation, especially during REM sleep, accompanied by a decreased amplitude of respiratory movements and endoesophageal pressure oscillations; (b) central apneas occurring usually (but not always) in REM sleep which are seldom accompanied by O_2 desaturation below 80%. When the degree of hypoventilation during sleep becomes severe, IPPV is recommended.

Bronchial Asthma

Many asthma sufferers tend to have dyspnoeic attacks during the night or in the early hours of the morning. This circadian trend seems to be closely linked to sleep since shift workers experience asthma attacks during daytime sleep. The relation between asthma and sleep is likely to be linked to the fact that there is a physiological increase in upper airway resistance, which peaks between 2 and 6 A.M., in all individuals and to a greater extent in asthma sufferers [56]. Asthma attacks do not coincide with specific sleep stages, but their relatively low incidence during sleep stages 3-4 could depend on the fact that these stages prevail in the early hours of sleep when upper airway resistence has not yet reached a peak. The mechanisms responsible for nocturnal narrowing of the upper airways remain unsettled. The fall in cortisol and epinephrine plasma levels does not seem to be relevant. However, the low levels of epinephrine during sleep may lead to local release of histamine by mast cells [57].

Nocturnal gastro-oesophageal reflux may be responsible for asthma attacks since reflux material from the stomach triggers reflex bronchoconstriction when it comes into contact with the oesophageal mucosa.

References

1. Burwell CS, Robin ED, Whaley RD, Bickelmann AG (1956) Extreme obesity associated with alveolar hypoventilation. A Pickwickian syndrome. Am J Med 21:811-818
2. Jung R, Kuhlo W (1965) Neurophysiological studies of abnormal night sleep and the Pickwickian syndrome. In: Akert K, Bally C, Schade JP (eds) Sleep mechanisms. Progress in brain research, vol. 18. Elsevier, Amsterdam, pp 140-160
3. Gastaut H, Tassinari CA, Duron B (1965) Etude polygraphique des manifestations épisodiques (hypniques et respiratoires), diurnes et nocturnes, du syndrome de Pickwick. Rev Neurol 112:568-579
4. Lugaresi E, Coccagna G, Berti Ceroni G (1968) Syndrome de Pickwick et syndrome d'hypoventilation alveolaire primaire. Acta Neurol Belg 68:15-25
5. Coccagna G, Mantovani M, Brignani F, Parchi C, Lugaresi E (1972) Continuous recording of the pulmonary and systemic arterial pressure during sleep in syndromes of hypersonnia with periodic breathing. Bull Physiopathol Respir 8:1159-1172
6. Lugaresi E, Coccagna G, Farneti P, Mantovani M,Cirignotta F (1975) Snoring. Electroencephalogr Clin Neurophysiol 39:59-64
7. Lugaresi E, Mondini S, Zucconi M, Montagna P, Cirignotta F (1983) Staging of heavy snorers'disease. A proposal. Bull Europ Physiopath Resp 19:590-594
8. Coccagna G, Cirignotta F, Lugaresi E (1991) Changes in general circulation in sleep apnea syndrome. In: Peter JH,Penzel T, Podszus T, Von Wichert P (eds) Sleep and health risk. Springer-Verlag, Berlin Heidelberg, pp 300-309
9. Cirignotta F, Zucconi M, Mondini S, Gerardi R, Lugaresi E (1989) Cerebral anoxic attacks in sleep apnea syndrome. Sleep 12:400-404
10. Coccagna G, Lugaresi E, Cirignotta F (1988) Sleep apnea syndrome and systemic hypertension. In: Duron B, Levi-Valensi P (eds) Sleep disorders and respiration, vol. 168. Colloque INSERM/ John Libbey Eurotext, London-Paris, pp 155-169
11. Smirne S, Iannaccone S, Ferini-Strambi L, Comola M, Colombo E, Nemni R (1991) Muscle fibre type and habitual snoring. Lancet 337:597-599

12. Kales A, Caldwell AB, Cadieux RJ et al. (1985) Severe obstructive sleep apnea. II: Associated psychopathology and psychosocial consequences. J Chron Dis 38:427-434
13. Lugaresi E, Cirignotta F, Coccagna G, Piana C (1980) Some epidemiological data on snoring and cardiocirculatory disturbances. Sleep 3:221-224
14. Koskenvuo M, Kaprio J, Partinen M, Langinvainio H, Sarna S, Heikkila K (1985) Snoring as a risk factor for hypertension and angina pectoris. Lancet 1:893-895
15. Norton PG, Dunn EV (1985) Snoring as a risk factor for disease: an epidemiological survey. BMJ 291:630-632
16. Gislason T, Aberg H, Taube A (1987) Snoring and systemic hypertension: an epidemiological study. Acta Med Scand 222:415-421
17. Schmidt-Nowara WW, Coultas DB, Wiggins C, Skipper BE, Samet JM (1990) Snoring in a Hispanic - American population. Risk factors and association with hypertension and other morbidity. Arch Intern Med 150:597-601
18. Telakivi T (1989) Breathing disturbance during sleep in adults. Clinical correlations in normal males, Down's syndrome and the dementias. Thesis pp 1-117
19. Hla KM, Young TB, Bidwell T, Palta M, Skatrud JB, Dempsey J (1994) Sleep apnea and hypertension. Ann Intern Med 120:382-388
20. Shepard JW (1986) Hemodynamics in obstructive sleep apnea. In: Fletcher EC (ed) Abnormalities of respiration during sleep. Grune and Stratton, Orlando, New York, S. Diego, pp 39-61
21. Kales A, Bixler EO, Cadieux RJ et al. (1984) Sleep apnoea in a hypertension population. Lancet 2:1005-1008
22. Lavie PV, Ben - Yosef R, Rubin AE (1984) Prevalence of sleep apnea syndrome among patients with essential hypertension. Am Heart J 108:373-376
23. Fletcher EC, De Behuke RD, Lovoi MS, Gorin A (1985) Undiagnosed sleep apnea in patients with essential hypertension. Ann Intern Med 103:190-195
24. Williams AJ, Houston D, Finberg S, Lam C, Kinney JL, Santiago S (1985) Sleep apnea syndrome and essential hypertension. Am J Cardiol 55:1019-1022
25. Coccagna G, Mantovani M, Brignani F, Parchi C, Lugaresi E (1972) Tracheostomy in hypersonnia with periodic breathing. Bull Physiopath Respir 8:1217-1227
26. Sullivan CE, Issa FG, Ellis E, Brunderer J, Mc Cauley Bye PT, Grunstein R, Costas L (1987) Treatment of cardiorespiratory disturbances during sleep. In: Von Hahn HP (ed) Interdisciplinary topics in gerontology, vol. 22. Karger, Basel, pp 47-67
27. Hung J, Whitford EG, Parsons RW, Hillman DR (1990) Association of sleep apnoea with myocardial infarction in in men. Lancet 336:261-264
28. Koskenvuo M, Partinen M, Kaprio J (1985) Snoring and disease. Ann Clin Res 17:247-251
29. Koskenvuo M, Kaprio J, Telakivi T, Partinen M, Heikkila K, Sarna S (1987) Snoring as a risk factor for ischaemic heart disease and stroke in men. BMJ 294:16-19
30. De Olazabal JR, Miller MJ, Cook WR, Mithoefer JC (1982) Disordered breathing and hypoxia during sleep in coronary artery disease. Chest 82:548-552
31. D'Alessandro R, Magelli C, Gamberini G, Bacchelli S, Cristina E, Magnani B, Lugaresi E (1990) Snoring every night as a risk factor for myocardial infarction: a case - control study. BMJ 300:1557-1558
32. Partinen M, Palomaki H (1985) Snoring and cerebral infarction. Lancet 2:1325-1326
33. Palomaki H, Partinen M, Juvela S, Kaste M (1989) Snoring as a risk factor for sleep related brain infarction. Stroke 20:1311-1315
34. Cirignotta F, D'Alessandro R, Partinen M et al. (1989) Prevalence of every night snoring and obstructive sleep apnoeas among 30-69-year-old men in Bologna, Italy. Acta Neurol Scand 79:366-372

35. Waller PC, Bhopal RS (1989) Is snoring a cause of vascular disease ? An epidemiological review. Lancet 1:143-146
36. Hoffstein V, Mateika S, Rubinstein I, Slutsky AS (1988) Determinants of blood pressure in snorers. Lancet 2:992-994
37. Hoffstein V (1996) Is snoring dangerous to your health? Sleep 19:506-516
38. Billiard M, Alperovitch A, Perot C, Jammes A (1987) Excessive daytime somnolence in young men: prevalence and contributing factors. Sleep 10:297-305
39. Lavie P (1983) Sleep apnea in industrial workers. In: Guilleminault C, Lugaresi E (eds). Sleep/wake disorders: natural history, epidemiology and long-term evolution. Raven, New York pp 127-135
40. Telakivi T, Partinen M, Koskenvuo M, Salmi T, Kaprio J (1987) Periodic breathing and hypoxia in snorers and controls. Validation of snoring history and association with blood pressure and obesity. Acta Neurol Scand 76:69-75
41. Gislason T, Almqvist M, Eriksson G, Taube A, Boman G (1988) Prevalence of sleep apnea syndrome among Swedish men - An epidemiological study. J Clin Epidemiol 41:571-576
42. Pasquali R, Colella P, Cirignotta F et al. (1990) Treatment of obese patients with obstructive sleep apnea syndrome (OSAS): effect of weight loss and interference of otorhinolaryngoiatric pathology. Internat J Obesity 14:207-217
43. Suratt PM, Mc Tier RF, Findley et al. (1992) Effects of very low caloric diet with weight loss on obstructive sleep apnea. Am J Clin Nutr 56:1825-45
44. Mondini S, Gerardi R, Pasquali R, Rinaldi-Ceroni A, Morselli P, Cirignotta F (1992) Effetti del dimagramento sulla sindrome dele apnee ostruttive nel sonno. In: Smirne S, Ferini-Strambi L, Zucconi M (eds) Il sonno in Italia. Poletto, Milano, pp 147-152
45. Sullivan CE, Issa FG, Berthon-Jones M, Eves L (1981) Reversal of obstructive sleep apnea by continuous positive airway pressure applied through the nares. Lancet 1:862-65
46. ASDA Standard of Practice Committee (1996) Practice parameters for the treatment of obstructive sleep apnea in adults: the efficacy of surgical modification of upper airway. Sleep 19:152-55
47. Sher AE, Schechtman KB, Piccirillo J (1996) The efficacy of surgical modification of the upper airway in adult with obstructive sleep apnea syndrome. Sleep 19: 156-77
48. Powell N, Riley R,Guilleminault C, Troell R (1996) A reversible uvulopalatal flap for snoring and sleep apnea syndrome. Sleep 19:593-599
49. ASDA Standard of Practice Committee (1995) Oral appliance for the treatment of snoring and obstructive sleep apnea: a review. Sleep 18:501-10
50. Cirignotta F, Mondini S, Gerardi E (1995) Benzodiazepine ed apnee notturne: è sempre un rischio? Rassegna di patologia dell'apparato respiratorio 5:457-459
51. Severinghans JW, Mitchell RA (1962) Ondine's curse. Failure of respiratory center automaticity while awake. Clin Res 10:122
52. Girandoux J (1939) Ondine. B Gresset (ed) Paris
53. Coccagna G, Lugaresi E (1978) Arterial blood gases and pulmonary and systemic arterial pressure during sleep in chronic obstructive pulmonary disease. Sleep 1:117-124
54. Douglas NJ, Calverley PMA, Leggett RJE, Brash HL, Flenley DC, Brezinova V (1979) Transient hypoxemia during sleep in chronic bronchitis and emphysema. Lancet 1:1-4
55. Coccagna G, Cirignotta F, Mondini S, Schiavina M, Gerardi R (1991) Sleep-related respiratory impairment in muscular and skeletal diseases. In: Peter JH,Penzel T, Podszus T, von Wichert P (eds) Sleep and health risk. Springer-Verlag, Berlin Heidelberg, pp 154-160
56. Hetzel MR, Clark TJH (1980) Comparison of normal and asthmatic circadian rhythms in peak expiratory flow rate. Thorax 35:732-738
57. Barnes P, Fitzgerald G, Brown M, Dollery C (1980) Nocturnal asthma and changes in circulating epinephrine and cortisol. N Engl J Med 303:263-267

Disorders of Motor Control and Sleep

P. Montagna, F. Provini, G. Plazzi and E. Lugaresi

Introduction

Sleep is physiologically characterized by a prominent reduction in muscular activity. Experimental experience with laboratory animals has helped define the relevant neural mechanisms underlying motor inhibition during the different stages of sleep. Major advances in the pathophysiology of motor control during sleep have been brought about by the widespread use of video-polysomnographic recordings in the clinical setting. Thus, the principal physiological motor accompaniments of sleep have been outlined in man, and several clinical conditions and nosological categories of motor disorders have been delineated. In this short survey of the main disorders of motor control arising during sleep we shall summarize our experience with patients presenting at the Centre for Sleep Disorders at the Institute of Neurology of the University of Bologna, where polysomnographic studies under audiovisual monitoring have been going on for the last thirty years. We shall deal in particular with the principal motor phenomena which are observed in normal humans during sleep: the so-called nocturnal myoclonus or periodic limb movements of sleep and the restless legs syndrome; nocturnal cramps; the recently-identified propriospinal myoclonus and its peculiar relation to the relaxation phase prior to sleep. The preceding disorders all occur during non-rapid eye movement (NRem) sleep stages. We shall also discuss the so-called Rem sleep behaviour disorders which are instead typical of rapid eye movement (Rem) sleep. More complex phenomena such as nocturnal epileptic seizures and parasomnias are beyond the scope of our survey.

Motor Control in Sleep

Both NRem and Rem sleep stages are characterized by motor quiescence and loss of muscle tone especially in antigravitary muscles, such as the chin (mental) muscles which are routinely monitored during polysomnographic (PSG) recordings. However, during Rem sleep, the prominent and universal muscular atonia is

Istituto di Neurologia, Università di Bologna, Via Ugo Foscolo 7, 40123 Bologna, Italy

interrupted by sudden, short-lived resumptions of motor activity, in the form of myoclonic jerks associated with the bursts of rapid eye movements which characterize the so-called phasic Rem (as opposed to tonic Rem) stages.

The mechanisms of muscle atonia during sleep have been elucidated in the experimental animal. The antigravitary muscles progressively relax in the transition from relaxed wakefulness to light sleep stages (stages 1 and 2 of NRem sleep), and atonia progresses again during the deep sleep stages (stages 2 and 3 of NRem sleep). Still, atonia does not become complete until Rem sleep. Prior, however, to the full development of atonia, i.e. in about the ten minutes of NRem sleep preceding the onset of the Rem stages, occasional episodes of complete loss of muscle tone occur [1]. The H-reflex, that is the monosynaptic reflex of the soleus muscle to electrical stimulation of Ia afferents in the tibial nerve, has been used as a measurement of motoneuron excitability in man during sleep. H-reflexes are progressively lost during the development of muscle atonia, reaching complete abolition during Rem sleep [1, 2], paralleling the evolution of the tendon jerks [3].

Such a tonic depression of motoneurons, maximal during Rem stages, is expressed in the inhibition of the recurrent alpha motoneuron discharges during sleep in the cat [4] and in the inhibition of the motor responses evoked by direct stimulation of motoneurons during sleep [5]. All these phenomena are indeed caused by a tonic decrease in excitability of ventral spinal motoneurons, which become hyperpolarized because of post-synaptic inhibition during Rem sleep [4-6]. These inhibitory influences in spinal motoneurons may be reproduced by direct stimulation of the inhibitory reticular formation [4, 5, 7]. It is thought that the somatomotor inhibition of sleep is due to the specific influence of neurons in the nucleus gigantocellularis of the reticular formation, in turn activated by pontomesencephalic neurons (nucleus pontis oralis) during Rem sleep. It is well known that lesions to these neurons cause suppression of the somatomotor inhibition typical of Rem sleep (so-called Rem-sleep without atonia [8, 9]) and produce striking behavioural abnormalities during which the animals exhibit complex motor activities like fighting, walking and searching behaviour in the absence of environmental awareness. Recent studies have defined distinct behavioural patterns and the neural mechanisms underlying these phenomena in the experimental animal [10, 11].

Atonia during sleep is not total, however, as PSG recordings on humans comprise a wide variety of bodily movements which normally occur and must be distinguished from the abnormal motor phenomena of sleep. Such physiological motor events of sleep are represented, in their most elemental form, by the so-called physiological hypnic myoclonus (PHM), first described clinically by De Lisi [12] in 1932 in both humans and animals. PHM consists of brief, irregular, and arrhythmic muscular twitches which usually involve the small muscles of the hand and face, sometimes in clusters and restricted to parts of bigger muscles, resembling fasciculation potentials. PHM has been studied quantitatively and shown to be particularly evident during Rem and NRem stage 1 sleep [13, 14], disappearing instead during deep sleep. Lasting less than 1 s, PHM is thought to represent activation of single or associated motor units by descending facilitatory volleys from the reticular formation. This hypothesis is consistent with the fact that PHM disappears on the affected side of patients with pyramidal or spinal lesions [13].

Sudden bodily jerks or sleep starts represent another motor accompaniment of sleep, first clinically described by Oswald [15] and widely observed in normal humans. They consist of sudden, jerking, involuntary movements which last up to 1 s and involve the whole or a part of the body, especially axial muscles. Sleep starts are typically observed during light sleep stages, and are associated with polygraphic signs of arousal (such as a K-complex on electroencephalogram (EEG), electrodermal activation and increased heart and breathing rates) and with peculiar feelings of "shock" and "falling into the void" (Fig. 1). Their mechanism and origin are fundamentally unknown, but they are often reported by patients and may cause problems in the differential diagnosis of seizures and epileptic myoclonus.

An even grosser motor accompaniment of normal sleep is represented by spontaneous bodily movements of sleep [16], which occur especially during the second half of the night prior to awakening and in fragmentary form, during light and Rem sleep stages. They are often preceded by signs of EEG arousals and are

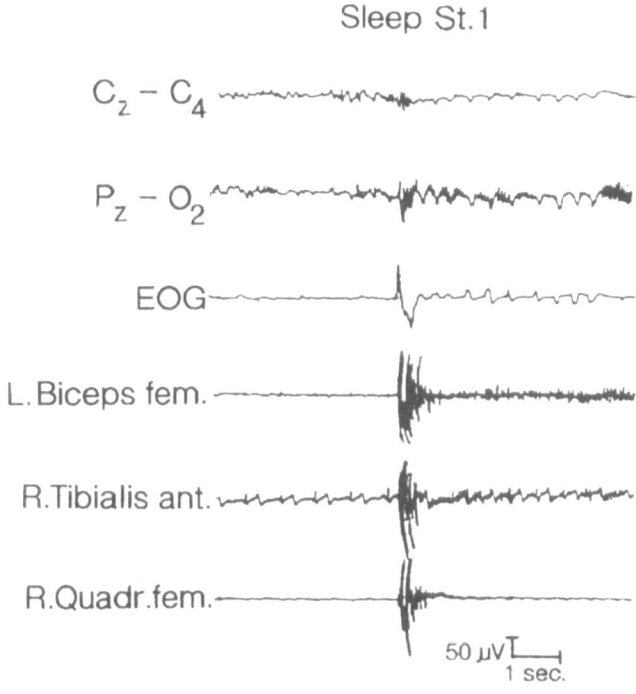

Fig. 1. *Polysomnogram of a patient with sleep starts.* Short (< 1 s), massive polyclonic activity appears at the transition from wake to sleep. Montage encloses EEG (C_z-C_4, P_z-O_2); electro-oculogram (*EOG*); left biceps femoralis electromyogram (*L. Biceps fem.*); right tibialis anterior electromyogram (*R. Tibialis ant.*); right quadriceps femoralis electromyogram (*R. Quadr. fem.*)

associated with changes of position during sleep. Spontaneous bodily movements are especially evident during infancy and appear to undergo a maturational process, as motor activity in adults progressively decreases and becomes periodic, reflecting the cyclic organization of sleep [17].

Abnormal Motor Phenomena of Sleep

Nocturnal Leg Cramps

Nocturnal leg cramps represent a disturbance of sleep which, even though often encountered in clinical practice, is still poorly characterized. This abnormal motor activity usually consists of painful contractions of the leg muscles, especially the small muscles of the sole of the foot or the sural muscles, which are rather sudden in onset and relieved by stretching the involved muscles. Muscle cramping is of course a well known condition in neuropathic disorders, in abnormal electrolyte conditions and in metabolic myopathies, but nocturnal leg cramping seems to differ from the former in that it occurs only during sleep, especially during NRem stages.

Recently, Jacobsen et al. [18] reported a familial disease of intermittent muscle cramps restricted to sleep, in which the cramps associated with myoclonic jerks involved the muscles of the trunk, limbs and face. Though the pathophysiology of the disease remained unknown, the cramps were found to respond to clonazepam. We have recently observed muscle cramps during sleep occurring as a familial condition and being responsive to carbamazepine [19]. Moreover, in the families that we reported, the disease was heterogeneous since electromyogram (EMG) and muscle biopsy signs of lower motor neuron involvement were variably associated.

In this regard, it is interesting to recall that infantile spinal muscular atrophy is characterized by spontaneous regular motor unit firing during sleep, especially stage 2 NRem sleep, which disappears during Rem sleep [20]. The origin of such pathological motor unit discharges is unknown.

Nocturnal Myoclonus and the Restless Legs Syndrome

Nocturnal myoclonus (NM) is a peculiar movement of sleep first described by Symonds in 1953 [21]. He reported patients complaining of sudden jerks arising during drowsiness and sleep but did not perform any neurophysiological investigation. Retrospectively, it is likely that most of his patients had physiological sleep starts. Symonds also thought that the jerks represented epileptic myoclonias, a view which was soon disputed. The correct interpretation and description of these sleep-related movements had to wait for the first polygraphic recordings, performed in Bologna, which showed these movements to consist of sudden dorsiflexion of the big toe and foot, often extending to the ankle, the knee and even the hip in the shape of a true triple flexion reflex. In extreme cases even the upper limbs may be involved with flexion at the wrist and forearm. The movements occur with patients usually unaware of them, during light sleep stages 1 and 2, progressively diminishing in deep sleep stages 3 and 4, and are usually completely suppressed during Rem sleep.This is not however an absolute finding, as in extreme cases they may occur even during Rem sleep, especially in a fragmentary form, and in particular during drowsiness, with the patient still not completely asleep. The last is especially true in severe cases of restless legs syndrome (RLS), and in such cases the movements may disturb the patient.

Duration of the jerks is usually between 0.5 and 5 s [22] and peaks at around 2 s, as measured from the duration of the EMG activity in the tibialis anterior muscle, the muscle routinely recorded during polysomnography. The jerks are not truly myoclonic according to standard conceptions, and may take different forms such as a simple tonic EMG activity, an isolated muscle jerk followed after a short while by tonic activity, or several jerks in a short sequence without tonic activity [23]. The jerks usually involve both legs, synchronously or not, but occasionally the jerks are restricted to one limb. In between the jerks, brief, isolated or repetitive arrhythmic EMG activity of lower amplitude has been noted [24], representing a continuous irregular firing of motor units. The most striking characteristic of nocturnal myoclonus is the repetition of the movements in a rhythmic or quasi-rhythmic fashion throughout the light sleep stages, for stretches of many minutes (Fig. 2). The interval between the jerks may vary widely from about 4 to 90 s, but has a mean of 20-40 s [22]. This feature which typifies NM has led to the term periodic limb movements in sleep (PLMS) being attributed to the phenomenon, emphasizing the fact that the jerks of NM may not be truly myoclonic in kind. Even the term PLMS does not account for the fact that sometimes NM arises during drowsiness and not sleep proper. In this short review the old term NM has been retained throughout.

Periodicity of NM has been related to the fact that the jerks quite often occur in association with K complexes on the EEG and with signs of autonomic activation (increased heart and breathing rates, increased blood pressure) which indicate arousal [23]. Light sleep is in effect a time of periodic arousals, displaying periodic oscillations in the EEG, autonomic functions and motor excitability. Such periodic oscillations occur at intervals of 20-40 s [23], a fact emphasized by Terzano et al. [25] who described the cyclic alternating pattern (CAP) during light sleep as consisting of phases of EEG activation alternating with epochs of inhibition. Thus NM should really be seen within the frame of these periodic oscillations of excitability of the nervous system during light sleep, probably modulated by some as yet unknown subcortical pacemaker.

Nocturnal myoclonus moreover may be seen as a paraphysiological phenomenon, since it is present in otherwise healthy people, especially elderly males. Studies of its prevalence in the general population document that NM is practically absent below the age of 30 years, but increases thereafter progressively especially after age 50 years, when up to 29% of normal people show NM [26, 27]. Therefore NM may be considered a physiological accompaniment of aging probably related to changes in sleep organization with age. It remains controversial whether NM should be considered a true cause of insomnia, a hypothesis endorsed by the current

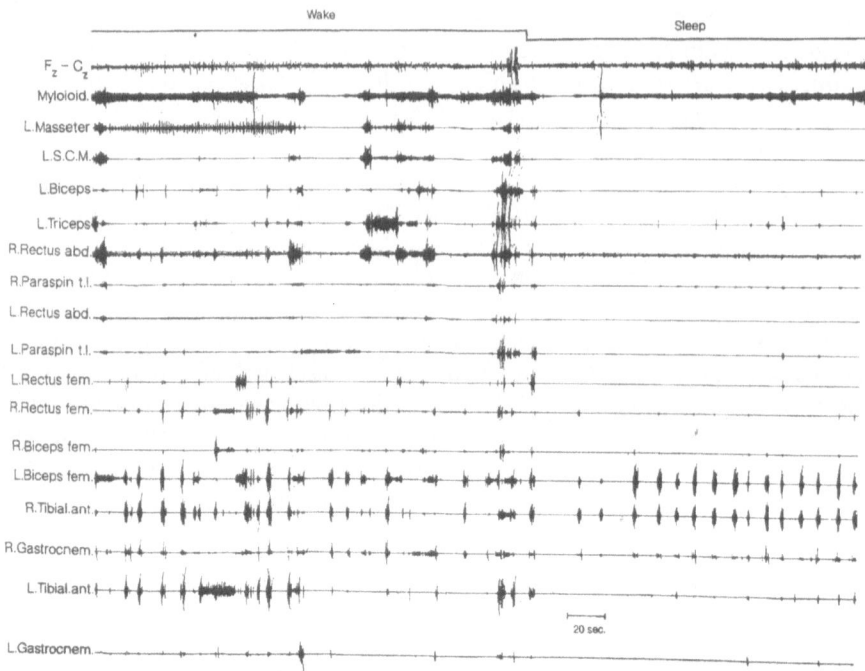

Fig. 2. *Polysomnogram of a patient with restless legs syndrome and nocturnal myoclonus.*
Voluntary movements and quasi-rhythmic polyclonic activities involving lower, upper
limbs and trunk when the patient lies in relaxed wake, that often increase at the transition
from wake to sleep, characterize restless legs syndrome. When the patient falls asleep the
periodical leg movements, typical of nocturnal myoclonus, appear. Montage encloses EEG
(F_z-C_z) and extensive electromyogram study of face, upper limbs, trunk, and lower limbs
muscles

"Associated Professional Sleep Societies" (APSS) classification of sleep disorders
[28]. Patients, in fact, do not usually complain of NM which is often a casual finding in polysomnographs, and the prevalence of NM has been reported to remain
the same among normal subjects and poor sleepers [29]. In our opinion, that NM
disrupts sleep causing insomnia may be a misunderstanding due to the fact that
NM is an almost invariable association of RLS which is the true cause of insomnia by way of the distressing paresthesia. Whether NM is causally related to several other neurological and non-neurological conditions "Amyotrophic Lateral
Sclerosis" (ALS), Isaac's and stiff person syndromes, peripheral neuropathies and
myelopathies also remains in doubt due to its high prevalence among the general
population. Since no controlled studies are available in any of these conditions,
their association with NM may represent simply a random event. NM is instead
significantly related to RLS, there hardly being any patient with this disease that
does not show NM at least once on repeated polysomnographies.

The origin of NM remains unknown. A subcortical site of origin, probably at
the level of the reticular substance [23], was postulated at first because of the

association of NM with autonomic and EEG oscillations. Electrophysiological investigations have recently excluded that the jerks arise at the level of the cerebral cortex, since back-averaging techniques failed to reveal any EEG activity preceding the abnormal movements (*Bereitschaftspotential*) [23, 30]. Reflexological studies of the mono- and polysynaptic relexes in RLS, a condition associated with NM, have shown increased segmental excitability at the spinal level suggesting modifications of the descending influences which modulate spinal cord excitability [31, 32].

We showed how NM can be induced and its periodicity modified by peripheral electrical stimuli at the level of the peroneal nerve [23], indicating that afferent sensory volleys from the periphery concur with central excitability changes during light sleep to generate NM. Recently, a detailed EMG study of NM jerks by Trenkwalder et al. [33] has shown that the EMG activity, though somewhat irregular from one jerk to the next, often sets on in the quadriceps muscle first and then propagates to other muscles of the lower limb according to a pattern of slow conduction along pathways intrinsic to the spinal cord, a feature typical of propriospinal myoclonus (see later). If confirmed, these findings imply that NM is generated at the spinal level. Supraspinal and peripheral influences probably act to release the phenomenon in the aged or in pathological conditions.

NM is significantly associated with RLS. The latter is characterized by distressing and ill-defined sensory disturbances, felt in the legs like an ache deep in the bone, a gnawing or burning sensation, as is usually felt below the knees. Associated with such disturbing paresthesia is an impelling need to move the limbs. Voluntary movements of the legs, like cycling, stretching or walking actually provide some temporary relief. The paresthesia is typically brought about by muscular rest and therefore characteristically sets on at bedtime when the patient tries to fall asleep, or in any other condition of muscular relaxation such as resting in an armchair or sofa. Patients while awake may make all kinds of fidgeting, dyskinetic movements of the legs, which become apparent on the EMG as continuous and irregular muscle activity of low amplitude [34]. The paresthesia and the associated urge to move severely disrupt sleep continuity, often forcing patients out of bed.

RLS is both an acquired and a genetic disorder. Acquired RLS is observed in the setting of myelopathies and peripheral neuropathies, such as polyneuropathy due to uremia or vitamin B_{12} deficiency, and may even herald the onset of familial amyloid polyneuropathy [35]. In about half of the cases, however, a familial recurrence of RLS is observed. Several pedigrees with RLS have been reported in which the disease is transmitted as an autosomal dominant trait [36-39]. In these families, symptoms of RLS usually appear in the second decade of life and persist with a fluctuating course for a lifetime [37]. In some families, NM has been reported as a isolated finding in siblings in the absence of the sensory disturbances typical of RLS [36]. Linkage studies in these families have up to now failed to disclose any significant association with genetic markers. Therefore pathogenesis of the disease is still unknown, though several hypotheses have been put forward. An abnormality in the metabolism of the endogenous opioids has been hypothesized, on the basis of the therapeutical efficacy of these substances [39]. Iannaccone et al. [40] instead, based on their findings from nerve biopsies documenting a loss of small calibre nerve fibres, believe RLS to be a polyneuropathy affecting small sensory axons. Finally, studies of patients with RLS by means of SPECT with IBZM (SPECT with IBZM), a ligand which selectively binds to D2 dopamine receptors, document reduced D2 dopamine transmission in the striatum in RLS [41]. None of these studies have been replicated and therefore the evidence remains inconclusive, or even controversial. For instance, functional MRI studies of the dyskinetic movements typical of RLS implicate the brain stem and red nucleus as sites of origin of the abnormal muscular activity [42]. It is likely that only molecular biology studies of familial RLS will be able to offer a verifiable insight into the pathogenesis of the disease.

Therapy of RLS has long been based on anecdotal reports, as befits a pathological condition of unknown origin. Recent years have, however, seen a more scientific approach based on controlled therapeutic trials. Thus treatment is based on the use of either benzodiazepines, in particular clonazepam, which is usually effective at doses of about 1 mg at bedtime [43], or levodopa, given at dosages of 250-500 mg in a single dose before going to sleep [44]. Both types of treatment have side effects and contraindications: clonazepam may sometimes provoke drowsiness the morning after, and may lose efficacy after some time of continued use. Levodopa, apart from the gastric side effects, has shown a worrisome tendency in some patients to increase RLS symptoms the day after, requiring an escalation of dosages and leading to a vicious circle of ever-worsening disability; the progressive increase in dosage should therefore be especially resisted. Other less used but sometimes effective medications are carbamazepine [45], baclofen [24, 46] and, in the case of severe cases not responding to any of the above medications, opioids [34]. While RLS is a pathological condition which usually requires the use of medications, NM, when an isolated finding, is usually asymptomatic and therefore does not require any treatment. Baclofen is however effective in curtailing it [46].

Excessive Fragmentary Myoclonus During NREM Sleep

Excessive fragmentary myoclonus during NREM Sleep is another form of abnormal motor activity during sleep which consists of discharges of brief myoclonic or fasciculation potentials, isolated or in short, repetitive bursts lasting less than 150 ms and occurring irregularly and asynchronously in different muscles throughout NRem sleep stages. This activity, found in different disorders such as sleep apnea and excessive daytime sleepiness, has been hypothesized to result from hypoxia during sleep [47].

Propriospinal Myoclonus

Myoclonus, defined as a sudden, shock-like muscular contraction, may originate in several sites of the central nervous system, among them the spinal cord. Frenken et al. [48] were the first to describe spinal myoclonus, occurring as quasi-rhythmic contractions of agonist and antagonist muscles innervated by one spinal segment. Such segmental myoclonus has been attributed to direct lesions of motor neurons [49] or to enhanced motor neuron excitability due to lesions of the inhibitory interneuronal networks which control motorneuronal excitability [50]. Spinal myoclonus has been studied during sleep, and shown to remain unaffected by the variations in excitability which characterize the different stages of sleep. Thus, it usually persists [51, 52] or ceases only intermittently during sleep [53]. This is in accordance with the evidence that spinal myoclonus may originate in a completely severed spinal cord [54] which is insensitive to the supraspinal influences associated with sleep.

More recently, however, another type of spinal myoclonus was described by Brown et al. [55] and termed propriospinal myoclonus (PSM), as it was thought to arise within the spinal cord and to diffuse to several muscular segments along propriospinal pathways intrinsic to the cord, sparing cranially-innervated muscles. Therefore, PSM is a plurisegmental myoclonus. In their first description of five cases, Brown et al. [55] reported myoclonus involving especially the axial muscles of the trunk, abdominal wall and hips, starting in one or a few spinal myotomes and propagating in an orderly fashion to more rostrally and caudally innervated muscles. Some cases had evidence for a cervical spinal lesion [56]. Later studies of other cases described a pattern of EMG bursts, each lasting 150-300 ms and occurring irregularly, spontaneously or sometimes evoked by somesthesic stimulations [57-59]. Detailed EMG studies document that the jerks originated first in the midthoracic segments and then propagated to rostral and caudal muscles at a velocity of about 3-11 m/s, suggestive of conduction along the intrinsic multisynaptic propriospinal pathways. In some cases, the first involved were the cervical segments. That the jerks of PSM were generated within the cord and were truly involuntary was shown by jerk-locked back-averaging techniques indicating the absence of the pre-movement *bereitschaftspotential* [55, 57] which characterizes myoclonic discharges originating in the cerebral cortex.

We observed four cases of PSM, in which the typical EMG bursts arose first in paraspinal or abdominal wall muscles, and then spread to rostral and caudal myotones at a velocity of 2-16 m/s, a pattern that is typical of PSM (Fig. 3). In our cases, however, we noticed that the jerks occurred spontaneously, sometimes in a quasi-rhythmic fashion, only when the patients were in a condition of relaxed wakefulness. The jerks were characterized by EEG alpha activity spreading to involve also the anterior brain regions (Fig. 4). This often happened when patients were left to themselves, for instance resting in an armchair or in bed in a supine position. That the supine position exerted a favouring effect in PSM was originally noted by Brown et al. [55]. We could, however, determine that it was not the bodily position itself, but rather the psycho-physiological condition of the subject that favoured the onset of the PSM jerks. The jerks of PSM were completely suppressed by sleep, even light sleep, disappearing as soon as theta activity was noted on the EEG, when patients became dozy or drowsy. Most relevantly, however, whenever patients were mentally stimulated, even when in a sitting or slouching position, and the alpha EEG activity disappeared, the PSM jerks also disappeared, thus confirming that PSM remained restricted to the peculiar psychic state of relaxed wakefulness with diffuse EEG alpha activity.

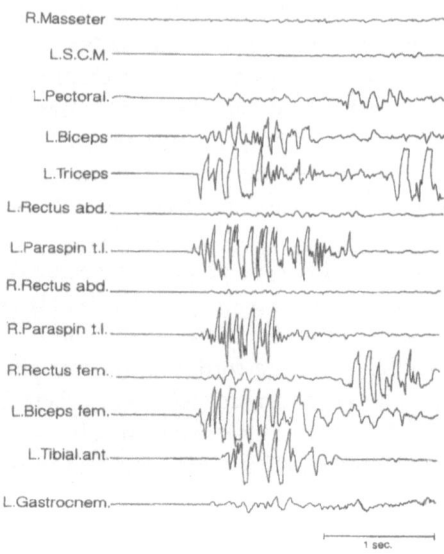

Fig. 3. *Extended wire electrode polygraphic study of a patient with propriospinal myoclonus.* The myoclonic activity arises in the left thoracolumbar paraspinal muscles (*L.Paraspin t.l.*) and spreads in muscles located rostrally and caudally. Delay to last activated muscle (*tibial. ant.*) is around 120 ms

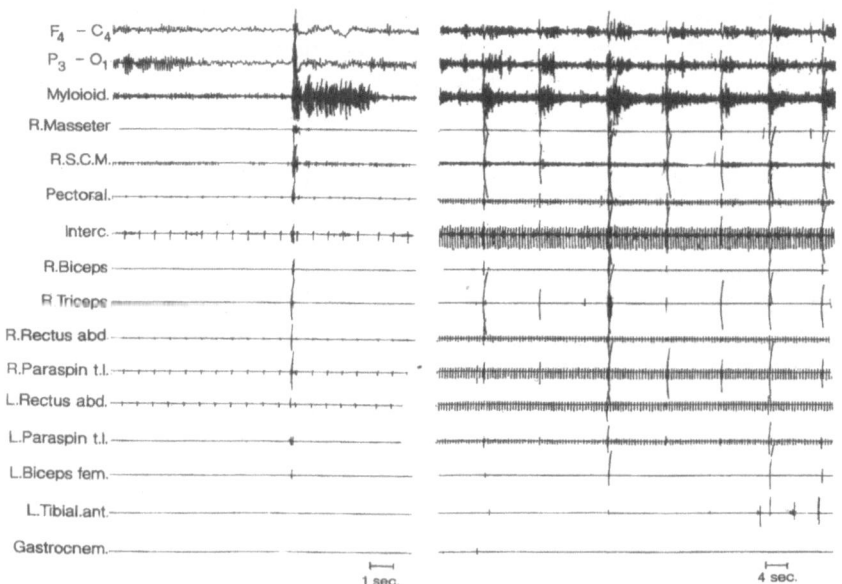

Fig. 4. *Polysomnogram of a patient with propriospinal myoclonus.* The *left side* (chart speed at 10 mm/s) of the figure illustrates the myoclonic activity arising at the vanishing of wake activity and the arousal reaction accompanying the jerk; the *right side* (chart speed at 2.5 mm/s) shows the periodicity of the myoclonus

This peculiar state-dependency of PSM is different from that of NM, which instead arises during relaxed wakefulness but then persists into light sleep. Our findings confirm that relaxed wakefulness is a peculiar physiological state with original psychic contents and mental mechanisms - the pre-dormitum state of Critchley [60], which likely act as supraspinal influences to facilitate the abnormal discharges originating in the spinal cord and causing the jerks of PSM.

REM Sleep Behaviour Disorders

REM sleep behaviour disorder (RBD) is a parasomnia affecting middle aged males characterized by intense motor or verbal paroxysmal dream-enacting episodes arising in REM sleep during loss of muscle atonia [61, 62] (Fig. 5). Clinical manifestations range from increased muscle twitching and jerks to complex, organized and finalistic motor and verbal activities leading to an enacted dream behaviour (Fig. 6). Such motor and verbal manifestations, as well as the content of dreams, are often fearful and violent, and can lead to physical injuries to partners and patients alike. When patients wake up during the episode they always recall a dream. The episodes usually appear at least one hour after falling asleep, coinciding with REM

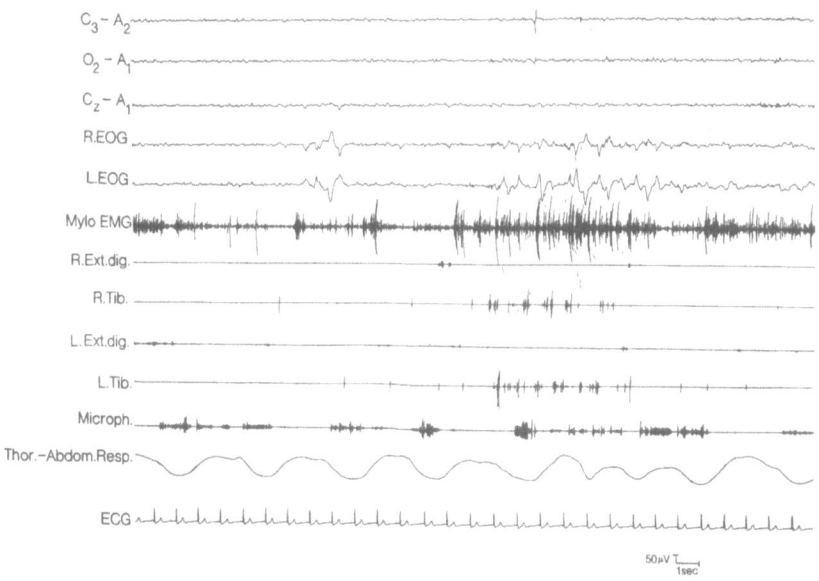

Fig. 5. *Polysomnogram of a patient with REM sleep behaviour disorder.* Montage required to document these paroxysmal episodes includes EEG (C_3-A_2, O_2-A_1, C_z-A_1), right and left electro-oculogram (*R.EOG, L.EOG*), chin electromyogram (*Mylo EMG*), right and left extensor digitorum (*R.Ext.dig* and *L.Ext.dig*) and right and left tibialis anterior electromyogram (*R.Tib.* and *L.Tib*). Increased muscle tone on chin electromyogram (*Mylo EMG*) is associated with excessive chin and limb myoclonic and polymyoclonic activity during REM sleep. Verbal activity is documented by the microphone (*Microph*). Autonomic activation is absent

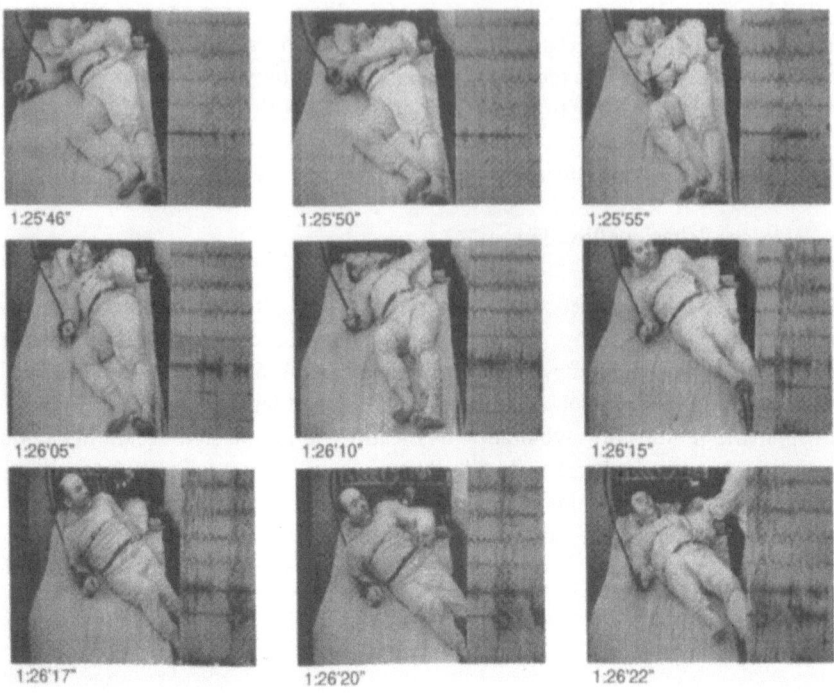

Fig. 6. *Rem sleep episode on video-polysomnography.* During REM sleep the patient displayed dream-enacting behavior: he screamed in dialect, tried to get out of bed terrified and acted defensively against the technician who approached him. Awakened he recalled that somebody was trying to shoot him. Episode duration: 1 min

sleep and may occur intermittently during the night. Due to the increase in REM sleep in the last third of the night, episodes are often more intense during the early morning hours, and are accompanied by the recall of vivid, fearful dreams. Episode frequency ranges from one or a few attacks per month to one or more every night. As in other REM sleep/parasomnias, autonomic activation is not dramatic.

Video-polysomnography is mandatory to confirm the diagnosis: extended PSG montages are required and the tracing has to be scored with allowance for intermittent or sustained loss of REM-atonia in chin, arm and leg muscles [63, 64]. RBD is more frequent in males; the mean age of onset ranges between 55 and 60 years of age. RBD can be idiopathic, but in approximately 60% of cases it is linked with neurological disease, or heralds a neurodegenerative disease. Structural lesions of the brain stem have been associated with RBD, with some studies emphasizing the role of vascular brain lesions [65]. MRI studies in these series disclose periventricular and pontine hyperintensities suggestive of small vessel arteriopathy, but the group of RBD with known etiology mostly includes narcolepsy and neurodegenerative diseases.

In their original series Schenck and Mahowald [61, 62] reported that 42.9% of RBD patients presented neurological diseases such as Parkinson's disease, Shy-

Drager syndrome, olivopontocerebellar atrophy, dementia, ischemic encephalopathy, alcoholism, multiple sclerosis, brain stem astrocytoma, Guillain-Barré syndrome and narcolepsy. In these patients RBD episodes usually responded readily to clonazepam 0.5-2 mg at bedtime, in both idiopathic and symptomatic cases. The follow-up study performed on the original patients with idiopathic RBD reported the appearance of a parkinsonian disorder in 38% of patients [66], demonstrating that the initial RBD diagnosed as idiopathic and responsive to clonazepam can precede the clinical onset of this neurodegenerative disease. The growing clinical relevance of RBD is emphasized by recent reports of RBD heralding extrapyramidal syndromes [67-70] or Lewy body disease [71]. A strong association of RBD with multiple system atrophy (MSA) has also been found [67-69, 72-74], even though RBD is not listed among the characteristic symptoms of MSA [75]. The short follow-up in the study of Schenck et al. [66] may have failed to solve the final diagnosis; the possibility of patients developing MSA was not ruled out.

In a recent study, after an adequate follow-up, we demonstrated that the vast majority of MSA patients had RBD [76]. We performed a clinical and PSG study on 39 consecutive MSA patients. Twenty-seven (69%) patients complained of nocturnal motor paroxysmal episodes related to dreams, suggesting the clinical diagnosis of RBD. In 12 (44%) of them RBD had preceded the clinical onset of the disease by more than one year. In seven (26%) RBD onset was concomitant with the appearance of motor or autonomic symptoms, and in eight (30%) it occurred at least two years after. On PSG recordings 35 (90%) of the 39 MSA patients had RBD. Our study established that RBD is a very common and important clinical sign in MSA, representing the most frequent clinical sleep manifestation and polysomnographic finding, and often heralding the appearance of other MSA symptoms by years.

HLA tissue typing performed on a sample of non-narcoleptic patients with RBD disclosed an association with specific HLA DQw1 class II genes (DQB1*05 and DQB1*06) [77].

Clonazepam is the drug of choice in RBD, and withdrawal of treatment usually leads to reappearance of attacks [62]. In a study performed on a few patients, alprazolam had overlapping efficacy with clonazepam [78].

We have seen how, in the 1960s, Jouvet and Delorme [8] documented the crucial role of tegmental pontine structures in the generation of abnormal REM sleep without atonia in the animal; this finding was confirmed by further lesional studies [9, 10]. Electrolytic lesions of the dorsal pontine tegmentum in the cat created REM sleep without muscle atonia and the animals displayed dream-enacting motor behaviour.

In humans, the nature of the mechanism producing RBD remains unknown, but the involvement of the REM sleep atonia cells of the pons is strongly suggested [65]. From this point of view MSA may represent a model disease: pathological and functional studies of this disease may shed light on the functional and anatomic alterations underlying RBD. In this regard, a significant loss of pontine neurons is often encountered in neuropathological examination of MSA patients [79].

Conclusions

In this short review we have described the main abnormalities of motor control in man during sleep, as they emerged through the routine use of videopolysomnographic methods in the last thirty years. Videopolysomnography has proved a most remarkable method in the delineation and full characterization of the features of these abnormal events of sleep, thus contributing substantially to their nosological categorization. Video-polysomnography in fact permits the actual monitoring of events which can only in part be reported by the subject asleep, or by the bed partner. Thus, several misconceptions have been corrected. For instance, the physiological bodily jerks of sleep or the nocturnal myoclonus, which, based on the purely clinical description, were initially thought of as epileptic in origin, have been more accurately described. Also, previously unknown events have been characterized, such as the Rem behaviour disorders, of which the patient is usually completely unaware.

Video-polysomnography is however a purely descriptive means and several questions remain unanswered regarding these abnormal motor events of sleep. Thus the origin and the anatomical site(s) involved in the pathogenesis of RLS and NM still escape us. On the other hand, new powerful imaging methods such as PET and functional MRI will undoubtedly clarify several issues in diseases such as PSM and NM, if the difficulties in application to a sleeping subject can be overcome. For the moment, the nosography of the abnormal motor events of sleep must per force be descriptive, based especially on the state-dependency of the events. Thus, motor events arising during the relaxation phase of wakefulness preceding sleep, the pre-dormitum of Critchley [60], behave independently and are different from those arising during NRem sleep proper, and Rem sleep stages. This state-dependency of the abnormal motor events of sleep has been the main frame of the classification followed here, and underlies the fact that the physiology of the motor system during each of these states is different, exerting differential effects on the various pathological underlying components.

References

1. Pivik T, Dement WC (1970) Phasic changes in muscular and reflex activity during non-REM sleep. Exp Neurol 27:115-124
2. Hodes R, Dement WC (1964) Depression of electrically induced reflexes ("H-reflexes") in man during low voltage EEG "sleep". Electroencephalogr Clin Neurophysiol 17:617-629
3. Kleitman N (1963) Sleep and wakefulness. University of Chicago Press, Chicago
4. Gassel MM, Marchiafava PL, Pompeiano D (1965) An analysis of the supraspinal influences acting on motoneurons during sleep in the unrestrained cat. Modification of the recurrent discharge of the alpha motoneurons during sleep. Arch Ital Biol 103:25-44
5. Morrison AR, Pompeiano O (1965) An analysis of the supraspinal influences acting on motoneurons during sleep in the unrestrained cat. Responses of the alpha motoneurons to direct electrical stimulation during sleep. Arch Ital Biol 103:497-516

6. Glenn LL, Foutz AS, Dement WC (1978) Membrane potential of spinal motoneurons during natural sleep in cats. Sleep 1:199-204
7. Moruzzi G (1972) The sleep-waking cycle. Ergeb Physiol 64:1-165
8. Jouvet M, Delorme F (1965) Locus coeruleus et sommeil paradoxal. CR Soc Biol 159:895-899
9. Morrison AR (1983) Paradoxical sleep and alert wakefulness: variations on a theme. In: Chase MH, Weitzman ED (eds) Sleep disorders: Basic and clinical research. Spectrum, New York, pp 95-122
10. Hendricks JC, Morrison AR, Mann GL (1982) Different behaviors during paradoxical sleep without atonia depend on pontine lesion site. Brain Res 239:81-105
11. Sanford LD, Morrison AR, Mann GL, Harris JS, Yoo L, Ross RJ (1994) Sleep patterning and behaviour in cats with pontine lesions creating Rem without atonia. Sleep Res 3:233-240
12. De Lisi L (1932) Su di un fenomeno motorio costante del sonno normale: le mioclonie ipniche fisiologiche. Riv Pat Ment 39:481-496
13. Dagnino N, Loeb C, Massazza G, Sacco G (1969) Hypnic physiological myoclonias in man: an EEG-EMG study in normals and neurological patients. Eur Neurol 2:47-58
14. Montagna P, Liguori R, Zucconi M, Sforza E, Lugaresi A, Cirignotta F, Lugaresi E (1988) Physiological hypnic myoclonus. Electroencephalogr Clin Neurophysiol 70:172-176
15. Oswald I (1959) Sudden bodily jerks on falling asleep. Brain 82:92-102
16. Gardner R, Grossman WI (1976) Normal motor patterns in sleep in man. In: Weitzman ED (ed) Advances in sleep research. Spectrum, New York, pp 67-197
17. Fukumoto M, Mochizuki N, Takeishi M, Nomura Y, Segawa M (1981) Studies of body movements during night sleep in infancy. Brain Dev 3:37-43
18. Jacobsen JH, Rosenberg RS, Huttenlocher PR, Spire JP (1986) Familial nocturnal cramping. Sleep 9:54-60
19. Marini C, Lia A, Liguori R, Monari L, Plazzi G, Tinuper P, Montagna P (1994) Familial cramping syndrome. It J Neurol Sci 7:64
20. Buchthal F, Olsen PZ (1970) Electromyography and muscle biopsy in infantile spinal muscular atrophy. Brain 93:15-30
21. Symonds CP (1953) Nocturnal myoclonus. J Neurol Neurosurg Psychiatry 16:166-171
22. Coleman RM (1982) Periodic movements in sleep (nocturnal myoclonus) and the restless legs syndrome. In: Guilleminault C (ed) Sleeping and waking disorders. Indications and techniques. Addison-Wesley, Menlo Park, pp 265-295
23. Lugaresi E, Cirignotta F, Coccagna G, and Montagna P (1986) Nocturnal myoclonus and restless legs syndrome. In: Fahn S et al. (eds) Myoclonus. Advances in neurology, vol. 43. Raven, New York, pp 295-307
24. Coccagna G, Lugaresi E, Tassinari CA, Ambrosetto C (1966) La sindrome delle gambe senza riposo (restless legs). Omnia Med Ther 44:619-687
25. Terzano MG, Mancia D, Salati MR, Costani G, Decembrino A, Parrino L (1985) The cyclic alternating pattern as a physiological component of normal NREM sleep. Sleep 8:137-145
26. Bixler EO, Kales A, Vela Bueno A, Jacoby JA, Scarone S, Soldatos CR (1982) Nocturnal myoclonus and nocturnal myoclonic activity in a normal population. Res Commun Chem Pathol Pharmacol 36:129-140
27. Coleman RM, Bliwise DL, Sajben N, De Bruyn L, Boomkamp A, Menn ME, Dement WC (1983) Epidemiology of periodic movements during sleep. In: Guilleminault C, Lugaresi E (eds) Sleep/wake disorders: natural history, epidemiology, and long-term evolution. Raven, New York, pp 217-229

28. Diagnostic Classification Steering Committee; Thorpy MJ Chairman (1990) International classification of sleep disorders: diagnostic and coding manual. Rochester, MN, American Sleep Disorders Association

29. Kales A, Bixler EO, Soldatos CR, Vela-Bueno A, Caldwell AB, Cadieux RJ (1982) Biopsychobehavioral correlates of insomnia, part 1: role of sleep apnea and nocturnal myoclonus. Psychosomatics 23:1-5

30. Trenkwalder C, Bucher SF, Oertel WH, Proeckl D, Plendl H, Paulus W (1993) Bereitschaftspotential in idiopathic and symptomatic restless legs syndrome. Electroencephalogr Clin Neurophysiol 89:95-103

31. Martinelli P, Coccagna G (1976) Rilievi neurofisiologici sulla sindrome delle gambe senza riposo. Riv Neurol 46:552-560

32. Wechslar LR, Stakes JW, Shahani BT, Busis NA (1986) Periodic leg movements of sleep (Nocturnal Myoclonus): an electrophysiological study. Ann Neurol 19:168-173

33. Trenkwalder C, Bucher SF, Oertel WH (1996) Electrophysiological pattern of involuntary limb movements in the restless legs syndrome. Muscle Nerve 19:155-162

34. Hening WA, Walters A, Kavey N, Gideo-Frank S, Cìté L, Fahn S (1986) Dyskinesias while awake and periodic movements in sleep in restless legs syndrome: treatment with opioids. Neurology 36:1363-1366

35. Salvi F, Montagna P, Plasmati R, Rubboli G, Cirignotta F, Veilleux M, Lugaresi E, Tassinari CA (1990) Restless legs syndrome and nocturnal myoclonus: initial clinical manifestation of Familial Amyloid Polyneuropathy. J Neurol Neurosurg Psychiatry 53(6):522-525

36. Boghen D, Peyronnard JM (1976) Myoclonus in familial restless legs syndrome. Arch Neurol 33:368-370

37. Montagna P, Coccagna G, Cirignotta F, Lugaresi E (1983) Familial restless legs syndrome: Long-term follow-up. In: Guilleminault C, Lugaresi E (eds) Sleep/wake disorders: natural history, epidemiology, and long-term evolution. Raven, New York, pp 231-235

38. Montplaisir J, Godbout R, Boghen D, De Champlain J, Young SN, Lapierre G (1985) Familial restless legs with periodic movements in sleep: electrophysiologic, biochemical and pharmacologic study. Neurology 35:130-134

39. Walters A, Hening W, Cìté L, Fahn S (1986) Dominantly inherited Restless Legs with myoclonus and periodic movements of sleep: a syndrome related to the endogenous opiates? In: Fahn S et al. (eds) Myoclonus. Advances in Neurology, vol. 43. Raven, New York, pp 309-319

40. Iannaccone S, Zucconi M, Marchettini P, Ferini-Strambi L, Nemni R, Quattrini A, Palazzi S, Lacerenza M, Formaglio F, Smirne S (1995) Evidence of peripheral axonal neuropathy in primary restless legs syndrome. Mov Disord 10:2-9

41. Staedt J, Stoppe G, Kogler A, Munz D, Riemann H, Einrich D, Ruther E (1993) Dopamine D2 receptor alteration in patients with periodic movements in sleep (nocturnal myoclonus). J Neural Transm Gen Sect 93:71-74

42. Seelos KC, Trenkwalder C, Bucher SF, Reiser M, Oertel WH (1996) High-resolution functional magnetic resonance imaging of involuntary limb movements in the restless legs syndrome. Neurology 46:A120

43. Montagna P, Sassoli de Bianchi L, Zucconi M, Cirignotta F, Lugaresi E (1984) Clonazepam and vibration in restless legs syndrome. Acta Neurol Scand 69:428-430

44. Brodeur C, Montplaisir J, Godbout R, Marinier R (1988) Treatment of restless legs syndrome and periodic movements during sleep with LDOPA: a double-blind controlled study. Neurology 38:1845-1848

45. Telstad W, Sorensen O, Larsen S, Lillwold PE, Stensrud P, Nyberg-Hansen R (1984) Treatment of the restless legs syndrome with carbamazepine: a double-blind study. BMJ 288:444-446

46. Guilleminault C, Flagg W (1984) Effect of baclofen on sleep-related periodic leg movements. Ann Neurol 15:234-239

47. Broughton R, Tolentino MA, Krelina M (1985) Excessive fragmentary myoclonus in NREM sleep: a report of 38 cases. Electroencephalogr Clin Neurophysiol 61:123-133

48. Frenken CWGM, Korten JJ, Gabreëls FJM, Joosten EMG (1974) Spinal myoclonus. Clin Neurol Neurosurg 77:44-53

49. Shivapour E, Teasdall RD (1980) Spinal myoclonus with vacuolar degeneration of anterior horn cells. Arch Neurol 37:451-453

50. Alvord EC, Fuortes MGF (1954) A comparison of generalized reflex myoclonic reactions elicitable in cats under chloralose anesthesia and under strychnine. Am J Physiol 176:253-261

51. Lugaresi E, Coccagna G, Mantovani M, Berti Ceroni G, Pazzaglia P, Tassinari CA (1970) The evolution of different types of myoclonus during sleep. Eur Neurol 4, 6:321-331

52. Bauleo S, De Mitri P, Coccagna G (1996) Evolution of segmental myoclonus during sleep: polygraphic study of two cases. It J Neurol Sci 17:227-232

53. Daniel DG, Webster DL (1984) Spinal segmental myoclonus. Arch Neurol 41:898-899

54. Bussel B, Roby-Brami A, Azouvi PH, Biraben A, Yakovleff A, Held JP (1988) Myoclonus in a patient with spinal cord transection. Brain 111:1235-1245

55. Brown P, Thompson PD, Rothwell JC, Day BL, Marsden CD (1991) Axial myoclonus of propriospinal origin. Brain 114:197-214

56. Brown P, Thompson PD, Rothwell JC, Day BL, Marsden CD (1991) Paroxysmal axial spasms of spinal origin. Mov Disord 6:43-48

57. Chokroverty S, Walters A, Zimmerman T, Picone A (1992) Propriospinal myoclonus: a neurophysiologic analysis. Neurology 42:1591-1595

58. Brown P (1994) Spinal myoclonus. In: Marsden CD, Fahn S (eds) Movement disorders 3. Butterworth/Heinemann, Oxford, pp 458-476

59. Schulze-Bonhage A, Knott H, Ferbert A (1996) Pure stimulus-sensitive truncal myoclonus of propriospinal origin. Mov Disord 11:87-90

60. Critchley M (1955) The pre-dormitum. Rev Neurol (Paris) 93:101-106

61. Schenck CH, Bundlie SR, Ettinger MG, Mahowald MW (1986) Chronic behavioral disorders of human REM sleep: a new category of parasomnia. Sleep 9:293-308

62. Schenck CH, Mahowald MW (1990) Polysomnographic, neurologic and psychiatric and clinical outcome report on 70 consecutive cases with REM sleep behavior disorder (RBD): sustained clonazepam efficacy in 89.5% of 57 treated patients. Cleve Clin J Med 57:10-24

63. Mahowald MW, Schenck CH (1994) REM sleep behavior disorder. In: Kryger MH, Roth T, Dement WC (eds) Principles and practice of sleep medicine, 2nd edn. WB Saunders, Philadelphia, pp 574-588

64. Lapierre O, Montplasir J (1992) Polysomnographic features of REM sleep behavior disorder: development of a scoring method. Neurology 42:1371-1374

65. Culebras A, Moore JT (1989) Magnetic resonance findings in REM sleep behavior disorder. Neurology 39:1519-1523

66. Schenck CH, Bundlie SR, Mahowald MW (1996) Delayed emergence of a parkinsonian disorder in 38% of 29 older men initially diagnosed with idiopathic rapid eye movement sleep behavior disorder. Neurology 46:388-393

67. Wright BA, Rosen JR, Buysse DJ, Reynolds CF, Zubenko GS (1990) Shy-Drager syndrome presenting as a REM behavioral disorder. J Geriatr Psychiatry Neurol 3:110-113

68. Culebras A (1992) Update on disorders of sleep and the sleep-wake cycle. Psychiatr Clin North Am 15:467-486

69. Tison F, Wenning GK, Quinn NP, Smith SJM (1995) REM sleep behaviour disorder as the presenting symptom of multiple system atrophy [letter]. J Neurol Neurosurg Psychiatry 58:379-380

70. Tan A, Salgado M, Fahn S (1996) Rapid eye movement sleep behavior disorder preceding Parkinson's disease with therapeutic response to levodopa. Mov Disord 11:214-216

71. Uchiyama M, Isse K, Tanaka K, Yokota N, Hamamoto M, Aida S, Ito Y, Yoshimura M, Okawa M (1995) Incidental Lewy body disease in a patient with REM sleep behavior disorder. Neurology 45:709-712

72. Coccagna G, Martinelli P, Zucconi M, Cirignotta F, Ambrosetto G (1985) Sleep-related respiratory and haemodynamic changes in Shy-Drager syndrome: a case report. J Neurol 232:310-313

73. Quera-Salva MA, Gulleminault C (1986) Olivopontocerebellar degeneration, abnormal sleep, and REM sleep without atonia. Neurology 36:576-577

74. Manni R, Morini R, Martignoni E, Pacchetti C, Miceli G, Tartara A (1993) A nocturnal sleep in multisystem atrophy with autonomic failure: polygraphic findings in ten patients. J Neurol 240:247-250

75. The Consensus Committee of the American Autonomic Society and the American Academy of Neurology (1996) Consensus statement on the definition of orthostatic hypotension, pure autonomic failure and multiple system atrophy. Neurology 46:1470

76. Plazzi G, Corsini R, Provini F, Pierangeli G, Martinelli P, Montagna P, Lugaresi E, Cortelli P (1997) REM sleep behavior disorders in multiple system atrophy. Neurology (in press)

77. Schenck CH, Garcia-Rill E, Segall M, Noreen H, Mahowald MW (1996) HLA class II genes associated with REM sleep behavior disorder. Ann Neurol 39:261-263

78. Schenck CH, Mahowald MW (1996) Long-term, nightly benzodiazepine treatment of injurious parasomnias and other disorders of disrupted nocturnal sleep in 170 adults. Am J Med 100:333-337

79. Daniel SE (1992) The neuropathology and neurochemistry of multiple system atrophy. In: Bannister R, Mathias CJ (eds) Autonomic failure. A textbook of clinical disorders of the nervous system, 3rd edn. Oxford University Press, London, pp 564-585

The Syndrome of Nocturnal Frontal Lobe Epilepsy

P. Tinuper, G. Plazzi, F. Provini, A. Cerullo and E. Lugaresi

Introduction

Partial epileptic seizures may arise in any point of the cerebral cortex. The clinical aspect of one seizure depends on the site of onset and the cortical area involved by the ictal discharge. Most of our knowledge of the semiological features of frontal partial seizures comes from stereotactic electroencephalogram (EEG) recordings of the seizures in patients undergoing functional neurosurgery for medically refractory partial epileptic seizures [1-9]. These studies describe attacks characterized by vocalization, screaming and complex motor behavior with bimanual and bipedal activity. Seizures are often nocturnal and tend to recur in clusters. Scalp EEG may be misleading in localizing the epileptic foci, in particular in seizures originating from the frontal lobes, due to the presence of large mesial and orbital surfaces [10].

The use of close-circuit video monitoring coupled with simultaneous EEG-polygraphic recording (video-EEG technique) has much improved the possibility to analyze the clinical and electrical features of paroxysmal events. Since its introduction, video-polygraphic monitoring has been widely used in sleep laboratories for polysomnographic studies of ictal events occurring during sleep.

In 1981 some of us [11] first described five patients with nocturnal episodes of dystonic-dyskinetic behavior. Due to the absence of clear-cut EEG interictal and ictal epileptic features, this condition was named hypnogenic paroxysmal dystonia, but an epileptic nature was not excluded, in particular for those patient who presented with repetitive, stereotyped short attacks [12]. Since then the epileptic or non-epileptic origin of these nocturnal attacks has been debated [13-18]. Finally, in 1990 we confirmed that the so-called nocturnal paroxysmal dystonia with short-lasting attacks is, in fact, a peculiar form of frontal lobe epilepsy [19].

During the last two decades, more than 120 patients were admitted to our Sleep or Epilepsy Centers for paroxysmal nocturnal episodes of probable epileptic nature. All patients had a complete clinical, neuroradiological and video-polysomnographic examination. In 95 patients we confirmed the diagnosis of frontal lobe seizures occurring during sleep. According to the different intensity,

Istituto di Neurologia, Università di Bologna, Via Ugo Foscolo 7, 40123 Bologna, Italy

duration and clinical features, we described three different types of ictal manifestations: nocturnal (typical) frontal lobe seizures (NFLE), epileptic arousals (EA), and epileptic nocturnal wanderings (ENW).

Nocturnal Frontal Lobe Seizures

These attacks are characterized by a sudden arousal from non-random eye movement (non-Rem) sleep, followed by a complex motor activity lasting several seconds and ending, without a confusional post-ictal phase, in resumption of sleep. Motor manifestations may consist in kicking or cycling activity of the four limbs, or rocking of the trunk, sometimes with semipurposeful repetitive movements mimicking sexual activity. Tonic asymmetric or dystonic postures are frequent; vocalization, screaming, or swearing may be present during seizures. Sometimes the motor activity may be very violent with the possibility of injury or falling out of bed. At the end of the episodes patients normally go back to sleep. If questioned, they may or may not recall a motor attack. Seizures may recur several times a night and normally are very frequent with rare nights free of seizures. The semiology of the seizures is very similar from one patient to another, and in each patient the episodes conserve a strictly individual stereotypy without changing during the night or from one night to another. In some patients attacks assume different intensity, representing fragments of the entire attack which may recur in a quasi-periodic sequence during a prolonged portion of sleep [19, 20].

Polygraphically, the seizures arise directly from non-Rem sleep; the onset of the attack coincides with an abrupt transition to wakefulness activity, often preceded by a K complex. EEG is then partially or completely masked by muscle artifactual activity, but in some cases a clear-cut low amplitude fast activity is recorded predominantly over the frontal regions; clear-cut spike-and-wave activity is rare. Tachycardia and breathing irregularities coexist during the seizure. Interic-tal EEG abnormalities are also rare, and sometimes evident only with sphenoidal electrodes.

Example 1

The patient was a 26-year-old, right-handed woman. Her 20-year-old sister presented partial and secondary generalized seizures (cryptogenic partial epilepsy) from 8 to 17 years of age and her mother suffered from sleep-walking episodes during childhood. Birth and development were normal. At age 8 nocturnal seizures began characterized by head rising and back arching, sudden complex motor automatisms involving both legs, arms and trunk, and bipedal movements similar to cycling accompanied by vocalization (Fig.1). Seizures last 30-40 s; at the end she sometimes would wake and not remember the episode. Neurological examination was normal.

Interictal EEG (Fig. 2) showed only a sharp theta activity on both frontal regions, more evident during REM sleep. Brain NMR showed a right cerebral hemiatrophy, a mild gliosis in the right frontal white matter and hypoplasia of the

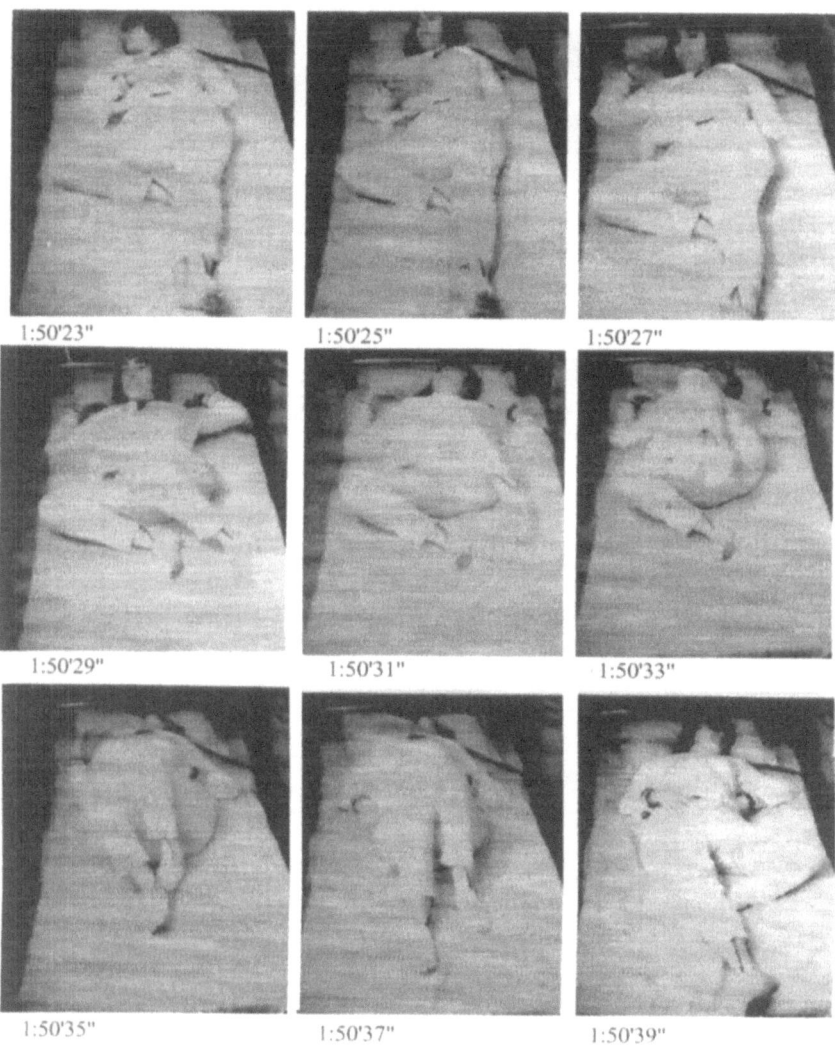

Fig. 1. *Nocturnal frontal lobe seizure recorded in example1*

right internal carotid artery. Ictal EEG (Fig. 3) was almost completely masked by muscle artifacts; sometimes a paroxysmal sharp wave activity was evident over the anterior regions. In the first year of illness she had nightly seizures, then antiepileptic treatment was started with phenobarbital 50 mg/day and carbamazepine 300 mg/day. She remained seizure-free to the age of 20 when seizures reappeared with the same semiology. At age 25, at our first observation, she had 10-15 seizures every night. This frequency remained unchanged during one year of follow-up. Gabapentin 900 mg/day and clobazam 10 mg/day added to carbamazepine were ineffective. She currently takes only carbamazepine 1000 mg/day.

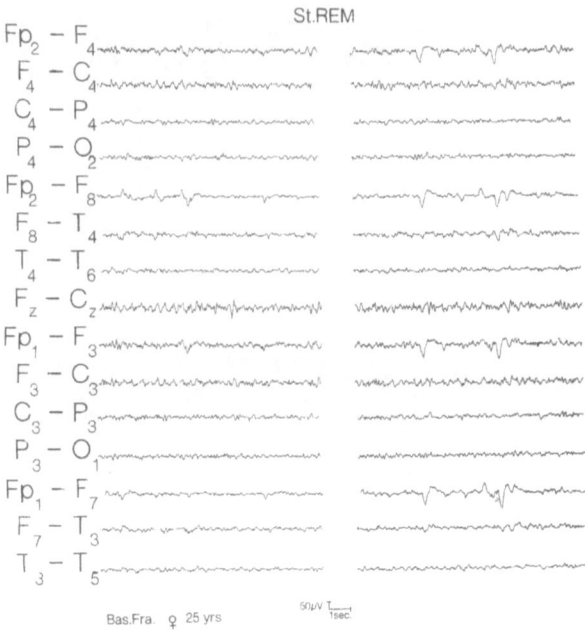

Fig. 2. *Interictal EEG.* During REM sleep, sharp theta activity is evident on both frontal regions and on the anterior vertex

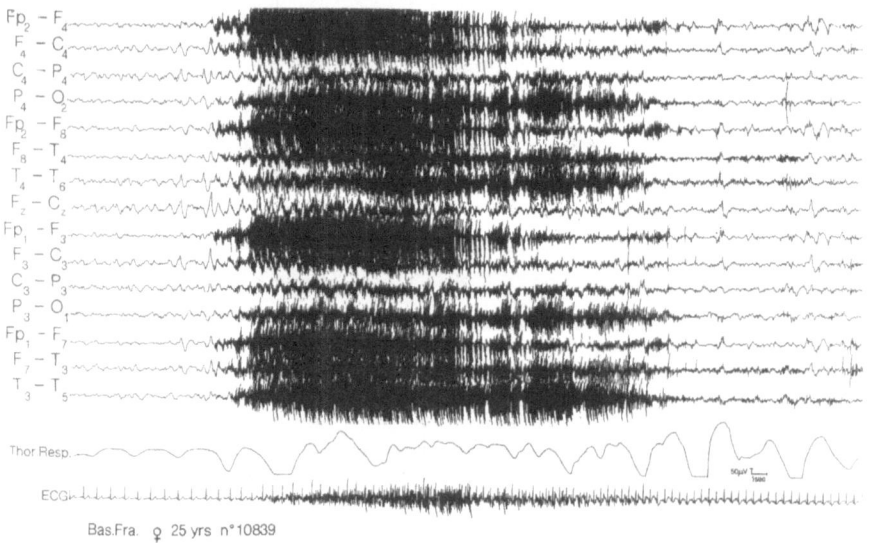

Fig. 3. *Ictal EEG recording.* During sleep stage 2, a seizure appears preceded by a diffuse sharp wave resembling a K complex. Paroxysmal EEG activity is then masked by muscle artifacts due to the seizure. No postictal abnormalities are evident. During the seizures tachycardia and breathing irregularities also occur

Epileptic Arousals

In 1990 [21], we described a first series of six patients presenting with repeated paroxysmal short-lasting arousals or awakening. Since then we have observed 37 more cases. We named this manifestation "paroxysmal arousal". The attacks are stereotyped in the same patient and, as in the cases of typical nocturnal frontal lobe seizures, may recur every night, sometimes several times a night.

Semiologically they consist in an abrupt awakening from non-Rem sleep; the patients raise one or both arms, sometimes raise their heads from the pillow, and open their eyes. Some slight dystonic posture or finger movements may occur. The episodes are very short (up to 3-5 s) and in 25% of the patients recurred periodically every 20-30 s for prolonged portions of non-Rem sleep. On EEG the episode is preceded by a K complex; the tracing is partially masked by muscle activity but sometimes it is possible to record a fast activity or a sharp-wave activity predominant over the frontal regions.

Example 2

The patient was a 14-year-old right-handed girl. Her father, mother and maternal uncle (II-3, II-2, II-1, respectively, Fig. 4) had somniloquy and somnambulism during childhood. Her two sisters (III-2, III-3) aged 15 and 20, respectively, presented nocturnal episodes with the same semiology as that of the proband. She was born after 6 months of pregnancy, but her development was normal. She suffered primary enuresis until 9 years of age. Since the age of 2 years she has presented nightly episodes lasting 1-2 min characterized by a sudden arousal with a frightened facial expression, in which she would scream and sit up in bed looking around; at the end she would go back to sleep (Fig.5). If awakened a few minutes later, she could not recall the episode. On rare occasions automatic ambulation with unintelligible mumbling occurred.

Neurological examination was normal as were interictal EEG during wakefulness and light sleep and nuclear magnetic resonance (MRI) of the brain. Ictal EEG showed paroxysmal diffuse sharp wave activity predominant anteriorly. She was not taking any therapy.

Epileptic Nocturnal Wanderings

Under the term epileptic nocturnal wanderings (ENW) we described episodes of paroxysmal ambulation and complex motor activity during sleep [22]. The episodes, recorded by video-polysomnography in four patients, were stereotyped and accompanied, on EEG, by a clear-cut epileptic discharge. These aspects differentiate these episodes from sleepwalking. In fact, somnambulism is a disorder of arousal [23,24], found in about 15-30% of healthy children, and characterized by semipurposeful behavior arising from non-Rem sleep. The possible epileptic

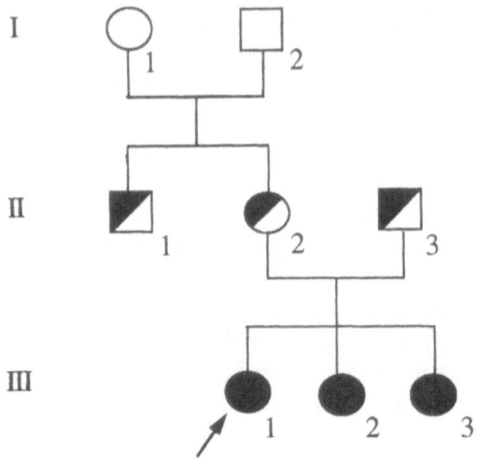

Somniloquy and Somnambulism

Paroxysmal arousal

Fig. 4. *Genealogic tree of example 2.* The patient (III-1) is indicated by *arrow*

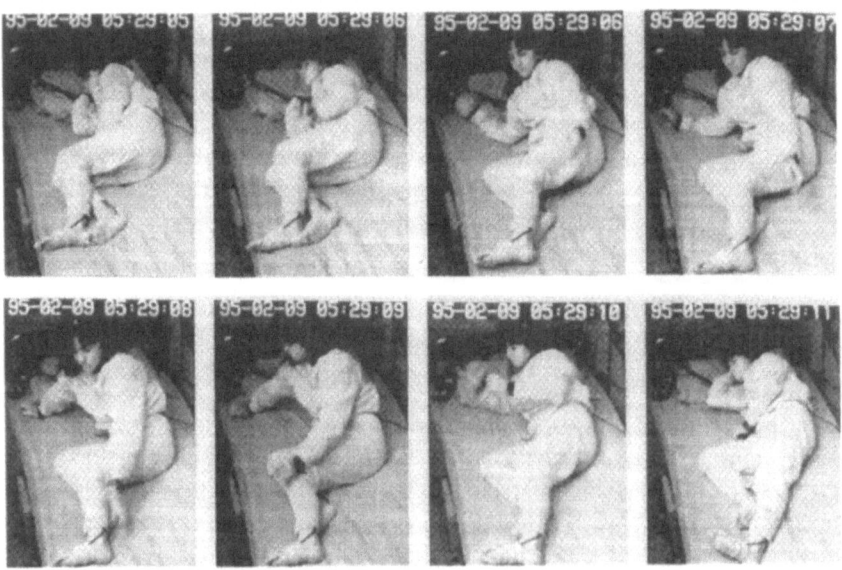

Fig. 5. *Epileptic arousal recorded in example 2*

nature of sleepwalking was ruled out by ictal EEG recording [23], but the true nature of these episodes, in particular when they responded to antiepileptic treatment, remained a matter of discussion [25-28]. Our patients presented with very frequent sleep episodes accompanied on EEG by a prolonged paroxysmal discharge over the anterior brain regions; interictal EEG abnormalities on the same regions were also present. In addition, all our patients with epileptic nocturnal wanderings presented typical nocturnal frontal lobe seizures.

Example 3

The patient was a 10-year-old boy with a positive family history for nocturnal paroxysmal episodes similar to the attacks displayed by the proband. His maternal grandfather and his mother had nocturnal attacks from infancy to 15 and 18 years of age, respectively; his 12-year-old sister has had the same type of episodes since age 5.

From age 3 the patient had nocturnal episodes arising during sleep: he would get up, walk around, open the door, step out into the garden and then go back to bed. Awakened by his parents, he never reported any dream content and in the morning he did not remember what happened during the night. From age 6 he has also presented several episodes during which he abruptly displayed violent movements and behavior. During these attacks he screamed and cried, got out of bed terrified and rushed around jumping. Then he went back to bed and fell asleep.

Neurological examination and enhanced brain computed tomography (CT) were normal. Brain MRI disclosed a small area of gray matter heterotopia in the right frontal lobe. Interictal EEG showed paroxysmal epileptic discharges over the right frontal region and on the anterior vertex. During all-night video-polygraphic monitoring we recorded several paroxysmal episodes with different severity. Some events consisted only of brief dorsal extensions of the right foot with a dystonic posture, tachypnea, eye-opening, a frightened expression, rubbing of eyes and then stomach. These minor episodes lasted a few seconds. On other occasions, the patient raised his head and trunk and stretched his legs out, rhythmically tapping the bed with his right hand; his neck was flexed in a dystonic posture with tonic stretching of the lips and buccal grimacing. On other occasions he jumped out of bed, with guttural vocalization, tried to open the door and started to jump around and on the bed, with arms stretched over his head, screaming and crying for his daddy (Fig. 6). With a terrified expression he would violently push away anybody trying to approach and restrain him. Questioned at the end of the episodes he could not report any dream and fell asleep right away.

These paroxysmal episodes would arise from any sleep phase, including REM sleep. On EEG, the minor attacks were characterized by spike discharges over both frontal regions, more evident on the right and the anterior vertex. The more prolonged and violent attacks resembling ENW were accompanied by sustained irregular spike-and-wave discharges with the same spatial distribution. Tachycardia accompanied the first part of the seizures (Fig. 7).

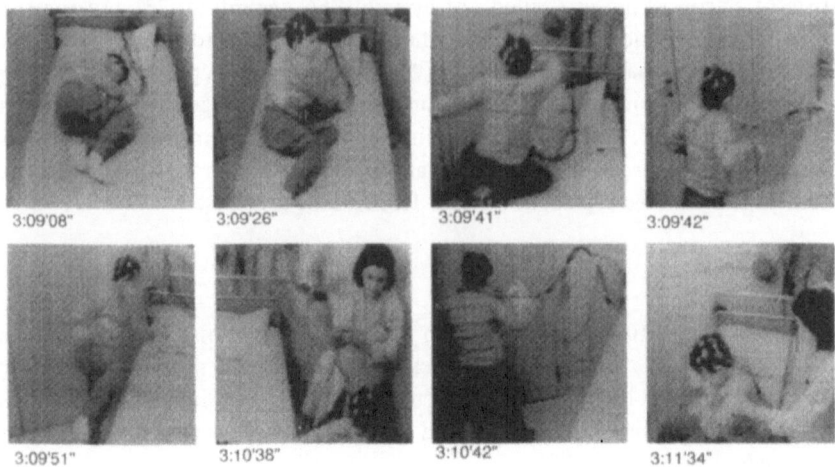

Fig. 6. *Episode of epileptic nocturnal wandering in example 3*

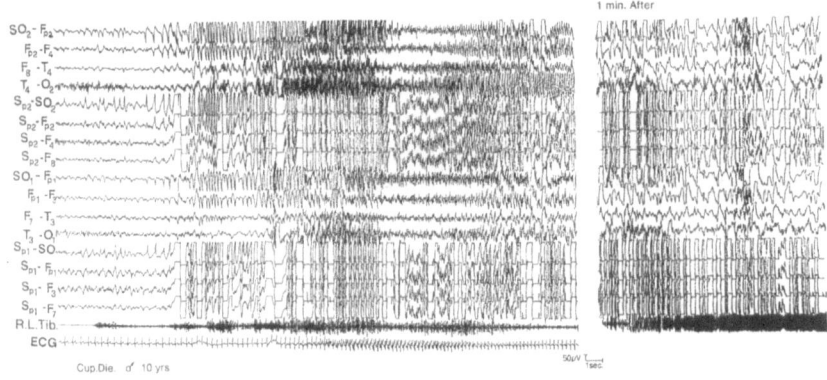

Fig. 7. *Ictal recording during a prolonged nocturnal attack shows a sustained irregular spike-and-wave discharge, diffuse on both hemispheres, with anterior predominance. SO_2, SO_1:* right and left supraorbital electrodes; *Sp_2, Sp_1:* right and left sphenoidal electrodes; *R.L.Tib.:* anterior tibialis muscles

Antiepileptic drug (AED) polytherapy (carbamazepine, phenobarbital, valproic acid, phenytoin, clobazam and lamotrigine) in different associations at therapeutic blood levels was ineffective.

During the all-night video-polygraphic recording, the patient's sister displayed several episodes arising from sleep: (a) brief repetitive events characterized by tentacular movements of the toes of both feet; (b) episodes during which she brought her left hand to her forehead, raised her head with eyes opened, and displayed chewing automatisms; (c) one full-blown attack during which she suddenly crossed her stretched-out legs, sat up in bed with a frightened expression, uttered incomprehensible words, rapidly moved her toes and fingers in a tentacular fashion, and then displayed chewing, gestual automatisms and unresponsiveness for two minutes. Most of the episodes began with a left frontal ictal epileptic discharge on the EEG, always associated with autonomic activation.

The Syndrome of Nocturnal Frontal Lobe Epilepsy

In our series of NFLE patients, 20% presented only typical frontal seizures and 9% only epileptic arousals, while no patient had only nocturnal wanderings. Concomitant episodes of NFLE, EA and ENW were present in 6%, and episodes of EA and NFLE, sometimes associated on the same night, were present in 34% of patients. These data confirm our hypothesis that nocturnal frontal lobe epilepsy is a spectrum of distinct phenomena, different in intensity, but representing a continuum of the same epileptic condition [29].

The physiopathological substrate leading a patient to have seizures only during sleep remains unclear: our hypothesis is that during non-Rem sleep the physiological events (K complexes) which periodically involve the frontal cortex activate an epileptogenic area in a predisposed subject [19]. This would explain the periodicity (every 20-40 s) of the paroxysmal events we observed in several patients [20] which is also documented in the normal sleep of healthy subjects [30].

NFLE affects both sexes with a slight prevalence for men (69%). The attacks begin more frequently between 14 and 28 years of age, but can affect any age. Seizures tend to increase in frequency during life, and this is a distinctive feature from parasomnias, which tend to disappear in the second decade of life. NFLE is frequently cryptogenic: in our population a known etiology was present in 13% of cases and neuroradiological studies disclosed abnormalities only in about one-third of patients.

In most cases ictal and interictal EEG did not show clear epileptiform abnormalities. However, a normal EEG does not exclude the diagnosis of NFLE. The origin of the epileptic disturbance in the deep frontal regions [6-8] makes it difficult for scalp EEG recording to detect paroxysmal abnormalities. In this case, the use of sphenoidal leads may be helpful.

Carbamazepine remains the first choice treatment, and is sometimes successful at very low doses taken in the evening. However one-third of the patients are resistent to all AED treatment.

Genetic Studies

Since our first observation [11] we have noted a strong familiar trait in our patients. In fact, about one-third of the patients had relatives with paroxysmal nocturnal episodes diagnosed as primary parasomnias. We confirmed this high percentage in all our further studies.

Recently, autosomal dominant nocturnal frontal lobe epilepsy (ADNFLE) has been defined as an epileptic syndrome characterized by short-lasting motor seizures with the features of frontal lobe origin, occurring in clusters mainly or exclusively during sleep and responsive to carbamazepine[31-33]. ADNFLE in a large Australian family has been mapped to chromosome 20, but heterogeneity is implied by the fact that not all families link to this chromosome [34,35]. Other authors [36] found, in a family with nocturnal frontal lobe epilepsy, a duplication of the centromeric heterochromatin region of chromosome 18. Recently [37] we tested the mutation CHRNA4 in DNA in a series of thirty NFLE patients (nine sporadic and twenty-one familial cases) and failed to find any abnormalities. Therefore, the genetic defects related to ADNFLE remain a matter for future studies.

References

1. Tharp BR (1972) A unique electroencephalographic and clinical syndrome. Epilepsia 13:627-642
2. Geier S, Bancaud J, Talairach J, Bonis A, Szikla G, Enjelvin M (1977) The seizures of frontal lobe epilepsy. Neurology 27:951-958
3. Wada JA, Purves SJ (1984) Oral and bimanual-bipedal activity as ictal manifestations of frontal lobe epilepsy. Epilepsia 25:668
4. Williamson PD, Spencer DD, Spencer SS, Novelly RA, Mattson RH (1985) Complex partial seizures of frontal lobe origin. Ann Neurol 18:497-504
5. Delgado-Esqueta AV, Swartz BE, Maldonado HM, Walsh GO, Rand RW, Halgren E (1987) Complex partial seizures of frontal lobe origin. In: Wieser HG, Engel J Jr (eds) Presurgical evaluation of epileptics. Springer-Verlag, Berlin, pp 267-299
6. Waterman K, Purves SJ, Kosaka B, Strauss E, Wada JA (1987) An epileptic syndrome caused by mesial frontal lobe seizure foci. Neurology 37:577-582
7. Morris HH, Dinner DS, Luders H, Wyllie E, Kramer R (1988) Supplementary motor seizures: clinical and electroencephalographic findings. Neurology 36:1075-1082
8. Wada JA (1988) Nocturnal recurrence of brief, intensely affective vocal and facial expression with powerful bimanual, bipedal, axial, and pelvic activity with rapid recovery as manifestations of mesial frontal lobe seizures. Epilepsia 29:209
9. Bancaud J, Talairach J (1992) Clinical semiology of frontal lobe seizures. In: Chauvel P, Delgado-Esqueta AV, Halgren E, Bancaud J (eds) Frontal lobe seizures and epilepsies. Advances in Neurology, vol. 57. Raven, New York, pp 3-58
10. Ajmone Marsan C (1988) Seizures originating from the orbital cortex of the frontal lobe. Epilepsia 29:208
11. Lugaresi E, Cirignotta F (1981) Hypnogenic paroxysmal dystonia: epileptic seizures or a new syndrome? Sleep 4:129-138
12. Lugaresi E, Cirignotta F, Montagna P (1986) Nocturnal paroxismal dystonia. J Neurol Neurosurg Psychiatry 49:375-380
13. Rajna P, Kundra O, Halasz P (1983) Vigilance level-dependent tonic sezures. Epilepsy or sleep disorder? A case report. Epilepsia 24:725-733
14. Crowell JA, Anders TF (1985) Hypnogenic paroxysmal dystonia. Acad Chil Psychiatry 24:353-358
15. Godbout R, Montplaisir J, Roleau I (1985) Hypnogenic paroxysmal dystonia: epilepsy or sleep disorder? A case report. Clin Electroencephalgr 16:136-142
16. Lee BI, Lesser RP, Pippenger CE, Morris HH, Luders H, Dinner DS, Corrie WS, Murphy WF(1985) Familial paroxysmal hypnogenic dystonia. Neurology 35:1357-1360

17. Berger HJC, Berendsen-VersteegTMC, Joosten EMG (1987) Nocturnal paroxysmal dystonia. J Neurol Neurosurg Psychiatry 50:647-648
18. Kovacevic-Ristanovic R, Golbin A, Cartwright R (1988) Nocturnal conversion disorder and nocturnal paroxysmal dystonia. Similiarities and treatments. Sleep Res 17:204.
19. Tinuper P, Cerullo A, Cirignotta F, Cortelli P, Lugaresi E, Montagna P (1990) Nocturnal paroxysmal dystonia with short-lasting attacks: three cases with evidence for an epileptic frontal lobe origin of seizures. Epilepsia 31:549-556
20. Sforza E, Montagna P, Rinaldi R, Tinuper P, Cerullo A, Cirignotta F, Lugaresi E (1993) Paroxysmal periodic motor attacks during sleep: clinical and polygraphic features. Electroencephalogr Clin Neurophysiol 86:161-166
21. Montagna P, Sforza E, Tinuper P, Cirignotta F, Lugaresi E (1990) Paroxysmal arousals during sleep. Neurology 40:1063-1066
22. Plazzi G, Tinuper P, Montagna P, Provini F, Lugaresi E (1995) Epileptic nocturnal wanderings. Sleep 18:749-756
23. Gastaut H, Broughton RJ (1965) A clinical and polygraphic study of episodic phenomena during sleep. In: Wortis J (ed) Recent advances in biology and psychiatry, vol. 7. Plenum, New York, pp 197-222
24. Broughton RJ (1968) Sleep disorders: disorders of arousal? Science 159:1070-1078
25. Tassinari CA, Mancia D, Dalla Bernardina B, Gastaut H (1972) Pavor nocturnus of non-epileptic nature in epileptic children. Electroencephalogr Clin Neurophysiol 33:603-607
26. Pedley TA, Guilleminault C (1977) Episodic nocturnal wandering responsive to anti-convulsant drug therapy. Ann Neurol 2:30-35
27. Maselli RA, Rosemberg RS, Spire JP (1988) Episodic nocturnal wandering in non-epileptic young patients. Sleep 11:156-161
28. Oswald J (1989) Episodic nocturnal wandering. Sleep 12:186-187
29. Montagna P (1992) Nocturnal paroxysmal dystonia and nocturnal wandering. Neurology 42(supp 6):61-67
30. Lugaresi E, Coccagna G, Mantovani M, Lebrun R (1972) Some periodic phenomena arising during drowsiness and sleep in man. Electroencephalogr Clin Neurophysiol 32:701-705
31. Scheffer IE, Bhathia KP, Lopes-Cendes I, Fish DR, Marsden CD, Andermann F (1994) Autosomal dominant frontal epilepsy misdiagnosed as sleep disorder. Lancet 343:515-517
32. Sheffer IE, Bhathia KP, Lopes-Cendes I, Fish DR, Marsden CD, Andermann E, Andermann F, Desbiens R, Keene D, Cendes F, Manson JI, Constantinou JEC, McIntosh A, Berkovic SF (1995) Autosomal dominant nocturnal frontal lobe epilepsy: a distinctive clinical disorder. Brain 118:61-73
33. Oldani A, Zucconi M, Ferini-Strambi L, Bizozzero D, Smirne S (1996) Autosomal dominant nocturnal frontal lobe epilepsy: electroclinical picture. Epilepsia 37:964-976
34. Phillips HA, Sheffer IE, Berkovic SF, Hollway GE, Sutherland GR, Mulley JC (1995) Localisation of gene for autosomal dominant nocturnal frontal lobe epilepsy to chromosome 20q13.2. Nat Genet 10:117-118
35. Berkovic SF, Phillips HA, Sheffer IE, Lopes-Cendes I, Bhatia KP, Fish DR, Marsden CD, Andermann E, Andermann F, Sutherland GR, Mulley JC (1995) Genetic heterogeneity in autosomal dominant nocturnal frontal lobe epilepsy. Epilepsia 36(suppl 4):147
36. Tinuper P, Montagna P, Cerullo A, Plazzi G, Provini F, Mochi M, Lugaresi E (1996) Autosomal dominant nocturnal epilepsy: a family with brain migration disorder and chromosome 18 duplication. Epilepsia 37(suppl 5):37
37. Mochi M, Provini F, Plazzi G, Corsini R, Tinuper P, Valentino ML, Lugaresi E, Montagna P (1997) Genetic heterogeneity in autosomal dominant nocturnal frontal lobe epilepsy. It J Neurol Sci (in press)

Subject Index